The Search for Physics.
Infinity.

The Search for Physics. Infinity.

Viktor Moroz

AuthorHouse™ LLC
1663 Liberty Drive
Bloomington, IN 47403
www.authorhouse.com
Phone: 1-800-839-8640

Title and illustration by Irina Alexeeva

Published by AuthorHouse 06/25/2014

ISBN: 978-1-4969-1990-8 (sc)
ISBN: 978-1-4969-1989-2 (e)

Library of Congress Control Number: 2014910967

Who Is This Book For?

Simple test: 2 + 2 = ?

A) How much do you wish? (Folklore).

B) And we sell or buy? (Folklore).

C) If at retail - a 5, if the gross - then 3. (Folklore).

D) 4

E) Sometimes 7, sometimes 8. (Folklore).

F) $2 + 2 \neq 4$ – Morris Kline, 1959, page 471: "Many educated people say today that 2 + 2 must be 4. But there are algebras, **physically useful algebras**, in which this statement **does not hold**."[1]

G) 2 + 2 = 4 – Morris Kline, 1980, page 215: " ...only **six thousand years** after ... the mathematicians could **finally prove** that 2 + 2 = 4."[2]

Case G) is illustration of The Second Law of the Research Institutions (scientific folklore):

The simplest task can be made arbitrarily complex.

This law works with The First Law of the Research Institutions (scientific folklore):

Research - is the satisfaction of personal curiosity at public expense.

The book is for who has open mind, and critical thinking,

Who get case D) for less then one second,

Who does not agree with cases F) and G), and

Who does accept that reference to traditions and/or authorities is logical fallacy of argumentation.

This book is dedicated to my mom, Anna Moroz.
She was physics' teacher, and took first class
in evacuation, 1941, in Siberia, Krasnoyarsk city.
18 students from her first class were going to be
physics' teachers.

And also book is dedicated to my wife, love and
inspiration - Irina Alexeeva.

Contents

Preface

About 300 years Physics was combination of mathematics, philosophy and experiments with observations. In modern time it is easy to see domination of mathematics over all other, and the fall of importance of experiments and observations from main role to auxiliary role, especially, to role of experiments to prove "experimentally" that 2 + 2 = 5. Herewith mathematics mutates from logic to powerful symbolism where can be accepted any concept, notion, axiom, by man wishes.

Doctor philosophy from 1957, Princeton University, Steven Weinberg, in his book *"Dreams of a Final Theory: The Search for the Fundamental Laws of Nature* (1993)," had written chapter VII "AGAINST PHILOSOPHY," to prove "the unreasonable ineffectiveness of philosophy." In chapter VI "BEAUTIFUL THEORIES" he provides point of view of Eugene Wigner, that mathematics has "unreasonable effectiveness."

The mathematician and historian of mathematics [3] - Morris Kline, had shown the illogical development of mathematics in his book "Mathematics. The Loss of Certainty," 1980 [2]. But in 1985, Morris Kline had done illogical step in his book "Mathematics and The Search for Knowledge" [4]: he tried to show a "truth" of illogical mathematics by a "useful" and "fruitfulness" application of mathematics in physics.

This work was done as attempting to move in logical direction from physics point of view, first of all, discussed the initial notions, and specially, notion of infinity. All facts and knowledge about inconsistency of infinity used in this work were and are well known, and were and are ignoring from logic point of view.

Inconsistency of notion of infinity well knows more than 2500 years. Usage an inconsistence notion in logical reasoning is logical mistake. But per Hermann Weyl, mathematics is science about infinity, per David Gilbert, usage infinity in set theory is mathematical paradise. Well known that mathematics pretends on status of the most logical science, and even more, per Emmanuel Kant, "In any particular theory there is only as much real science as there is mathematics."

Usage notion of infinity in mathematics, and usage mathematics in physics raises a question: when does mathematics itself become a science, and what is the science? The worth of mathematics is its deductive nature, and this means that "in mathematics there is only as much real science as there is logic," and this formula applicable to science too.

There is problem with logic too – man is so weak to follow logic, and man feels himself so powerful to modify logic per his wishes. Man prefers beauty and intuition, but not logic, and most of all man loves paradoxes. Smart man prefers point of view of the well known authority, especially authority which sits on the money.

Well known that a reference to authority, and to tradition, is logical fallacy of argumentation. In this work reference to authority will use to show this renown of facts, concepts, and notions – but not to argue its correctness or truth.

Objectives this book is attempting to follow logic way in physics, first of all, and in mathematics as tool for physics. There are many historians of mathematics [5-7], but one of them – Morris Kline – shown illogical development of mathematics. Illogicality of mathematics was found in paradoxes followed from its axioms, but inconsistency of mathematics begins in so called "undefined notion" and in axioms itself, axiom which used undefined notion. It is vicious circle – attempt "solve" paradox without analyzing of undefined notion and axiom itself. Understanding inconsistency of used notion and axiom follows to conclusion that we have approximate model or theory which has limited features for extrapolation and interpolation.

Using terms "undefined notion" and "axiom" implies accepting those terms without analyzing them for inconsistency, but we should do that because all notions have its base - our experience, direct or indirect, implicit or explicit, for who believe it or not, - and our sense and even mathematical intuition can make mistakes. Main idea this book is attempting to show problem in mathematics and, especially in physics, - wrong way of usage a symbolic theory:

Start from real object and/or phenomenon;

Make abstraction and express it in symbolic form;

Do idealization with inconsistency;

Inconsistency of ideal abstract form assigns to real object;

Relation to the start real object is used as argument of existence of the ideal object;

Transfer to physics a math method to create abstract object by declaration properties this object, and then "create new" real object;

"Discover" the real object which is necessary to "solve" the inconsistency of the ideal abstraction.

It is old known problem, which is completely ignoring in physics.

Lack of reflection in math show us that mathematicians "invented" methods to hide – not solve – this problem. There are: acceptance so called "undefined" notions, axiomatic method, acceptance conflicting sets of axioms, and so on.

Lack of reflection in physics brings the complete faith to math; loosing physical objects, transfer a inconsistency of man's theory to reality, declaration and the faith that Nature follows the math laws, experimental confirmation of existence of a non-existent.

This work is attempting to separate the abstract symbol world from reality, the faith from logics, the science from religion activity – it is, in short words, meaning of The Search For Physics, and the notion of Infinity is "undefined" notion which is used in mathematics and physics.

Introduction

Fact is that all or almost all science had a philosophical-religious ground. The philosophy is not science because it studies everything. The religion is the faith in Authority. A science is a part of faith, and part of logics: more faith shows religious activity, more logics gives scientific activity.

Who is older – philosophy or mathematics – it is hidden in a deep of history. It is known that Schumer civilization had a counting before a writing for 6 thousand years ago, and had known that the Earth goes round the Sun - for thousands years before Copernicus [8]. Per Aristotle and encyclopedia ancient meaning Physics was natural philosophy.

When the Earth was warm and Dinosaurs and Mammonthus wander on its surface, and … But it is too long story.

In short words, why The Search for Physics and why Infinity do you see on title page?

My mom was the physics' teacher, and my, if not first, then second book was notebook of physics, and I read a lot books about physics and physicists. Almost whole my life I knew that Faraday was one of the greatest physicist, pure physicist, because he worked with physical objects: wires, magnets, candles, and so on.

One fine day, I found that Faraday hate when somebody called him "physicist", he called himself as non-mathematical philosopher [9].

Well, what we have: Newton was the mathematical natural philosopher, Faraday was the non-mathematical philosopher, Maxwell translated Faraday's works into mathematical language, contemporary physicist had postgraduate academic degree as Doctor of Philosophy, and now, actually, what is physics?

But it was the smaller part of the question.

One another fine day, in 2011, I accidentally had bought book by Morris Kline, "Mathematics. The Loss of Certainty", 1980. In the shortest time I had swallowed it, I was shocked, and got afflatus of new understanding a lot thing, and that it was the main book in my life.

I guess that still did not estimate his book neither Morris Kline nor contemporaries.

Morris Kline did not understand value of his book, because he had written, in 1985, next book "Mathematics and The Search for Knowledge".

First mention book has four chapters, from V to VIII, which begins by words:

"The Illogical Development … ", and all of that about mathematics, about an illogical logic of mathematics.

Next Kline's book, on our point of view, was written in illogical direction.

It was attempting justify illogicality of a mathematical theory by the fruitfulness usage of math in physics.

But it is very well known, that, if we find illogicality, then neither next step nor start assumption cannot be accepted in logical reasoning.

Start step in logical reasoning we can see in Kline's books: there are so called "undefined notions". Morris Kline had seen the illogical development of mathematics in known paradoxes, but problem was and is in illogicality, in inconsistencies of "undefined notions", it is source of paradoxes.

Obvious, next step is necessity of review the physical theories with respect to illogicality of mathematics. Well known that mathematics is science about infinity, and INFINITY is the main initial and "undefined" notion in mathematics and physics.

This book is attempting to discuss several questions in logical, from our point of view, direction.

Book consists two parts: I. INFINITY, and II. THE SEARCH FOR PHYSICS.

Part I. INFINITY is mathematical part where we will discuss undefined or initial notion, notion of infinity and derived notions.

1. The Begin from the Beginning. This section dedicate importance of undefined notion which should call as initial notion and they should be analyzes on inconsistency.
2. World Well Known Notion of Infinity. Describe inconsistency of notion of infinity and problems which associated with reference on infinity. Emphasizes illogicality statements like "infinite number," "all number"
3. Set Theory. Shows problems with reference on infinity in set theory, and with one-to-one correspondence and illogicality of statement about equivalency part and whole of infinity set.
4. Signs of Infinity. Presents main symbols of infinity, and underlined that usage of symbol of infinity does not eliminate inconsistency of infinity.
5. Algebra of Infinity. Presents algebra with the infinity's symbols, different and similar to algebra of numbers.
6. Sergeyev's Grossone Theory. Examines inconsistency of the newest theory of infinity.
7. Usage of infinite expressions: 1 ? 0.999 … Provides detail analysis of usage of the infinite expressions on example of comparison unit and 0.999 … Shows that we have more reason to accept that 1>0.999 … than 1=0.999 …, but logical conclusion is rejection of usage of the infinite expressions.
8. Geometry – usage rough physical models. Emphasizes inside contradiction of initial notions of the Euclid's and non-Euclid's geometries.

9. Paradoxes: Banach – Tarski: paradox, theorem. Describes dependency of paradoxes from initial notions. Banach-Tarski paradox proves inconsistency the infinite division of a finite geometrical object.
10. Mathematics and Physics. Reminds that mathematics has more sources of inconsistency than infinity and derived from it notions, and all of them affect physics.
11. Logic and Logics. Discusses problems with logics which can be illogical too.
12. Math method: abstraction, idealization, generalization. Explains power and weakness abstraction and idealization as sources of inconsistency and illogicality.
13. Physical "axiom" – reality of physical object. Provides point of view on physical approach as opposed to mathematical methods.

Part II. THE SEARCH FOR PHYSICS. Present discussion of the physical initial notions and steps of the physical approach of study Nature.

14. What is Physics? Remind standard definitions of physics and shows open questions about the core of physics.
15. Substance – space – motion. Discusses initial notions of physics.
16. Dimensions. Explains dimension as next level of abstraction of moving in space substance.
17. Mass. Describes problems with concept of mass and diversity of meaning this notion.
18. Time. Shows physical approach to notion of time as artificial abstract notion from standard uniform moving body, that it has the same level of abstraction as dimensions, but we cannot mix them in space-time.

19. Maxwell's equations. Explains problems with Maxwell's equation from physical and mathematical point of view.
20. Light. Argues that we have still open question: what is light?
21. Electrical Charges – surface phenomena. Discusses problems with notion of the electric charges.
22. Electron. Shows problem with notion of electron and that we still have open question: what is electro?
23. SRT, GRT. Addresses to problems with initial concepts of these theories.
24. Physical method. Selects main aspects of the physical approaches.

Conclusions. We have reality and symbolic theories. Concepts these theories are abstract notion associated with reality, and main problems of mathematics and physics are the usage inconsistent notions and the reviving of abstract notions.

Appendix. Review of the book's topics per web site edge.com. Briefly shows topicality of the book's subjects.

To see a World in a grain of sand,
And a Heaven in a wild flower,
Hold Infinity in the palm of your hand,
And Eternity in an hour.

Auguries of Innocence (1803),
William Blake (1757 – 1827).

Part 1. Infinity.

There is well known opinion that mathematics is used to study Nature. In [1, page vii] Morris Kline write:

"Insofar as it [math] is a study of space and quantity, mathematics directly supplies information about these aspects of the physical world."

Morris Kline as well shows illogicality of mathematics [2]. Here we should notice two issues:

First of all, there is declaration of necessity of undefined notions.

And, secondly, there is consideration of illogicality of math as existing of paradoxes only, without analyzing of the initial notion, called as undefined notion.

Arguments for usage of the undefined notion were and are that:

1) We cannot infinitely define one notion through other;
2) Undefined notion is in reality initial notion which has clear meaning by mathematical or Descartes' intuition;
3) Undefined notion will "define" in axiom where this notion will use.

Infinity is one of initial and very important for mathematics notion. Herman Weyl defined mathematics as science about infinity.

Notion Infinity produced other notions: infinitesimal, continuity, infinite set and so on. Many branch of mathematics use notion of infinity and derived from it concepts. Let's begin from the beginning and consider term of "undefined" notion.

1. The Begin from the Beginning.

The Begin of the Beginning is "undefined notions," which we implicitly or explicitly defined from everyday experience, and we should analyze those notions on inconsistency. Morris Kline had written [2], that mathematics needs undefined notion, because we can not infinitely build definition with reference on other notion. But, from physics point of view, finite environment gives us finite start: the beginning of notion is a correspondence of a notion with a property of physical object. "Infinitely build definition with reference on other notion" directly show us that mathematics has symbolic nature. Not only math branches have initial notions and concepts, all science have them too, but initial notions can not be "undefined" to be used. It looks like usage word "undefined" needs to be free from spends time to analyze meaning and context of applicability this notion.

First of all, it is not make sense word "undefined" – any word at least defined by our everyday's experience or intuition which based on everyday's experience. Actually, we can generate some word as chain of characters (for example, by using program GUIDGEN.exe), but we cannot use it without establishing of a correspondence with some kind of environment: with software as ID or password, or with other words.

The necessity of undefined term well known a long time [2, page 24]:

"By Aristotle's time the requirement was certainty in effect, for he is explicit about standards of rigor such as need for undefined term and laws of reasoning."

And further [2, page 120]:

"Now Aristotle had pointed out that a definition must describe the concept being defined in terms of other concepts already known. Since one must start somewhere, there must be, he asserted, undefined concepts to begin with. Though there are many indications that Euclid, who lived and worked in Alexandria about 300 BC, knew the works of the classical Greeks and of Aristotle in particular, **he nevertheless defined all his concepts.**

There have been two explanations of this flaw. Either Euclid did not agree that there must be undefined terms or, as some defenders of Euclid state, he realized that there must be undefined terms but **intended his initial definitions to give only some intuitive ideas of what the defined terms meant**, so that one could know whether the axioms which were to follow were indeed correct assertions. In the latter case he should not have included the definitions in his text proper. Whatever Euclid's intentions were, practically all the mathematicians who followed him for two thousand years failed to note the need for undefined terms. Pascal in his *Treatise on the Geometrical Spirit* (1658) called attention to this need but his reminder was ignored."

[2, page 216] "However, we may recall (Chapter V) that Euclid's **definitions of line and other concepts were superfluous.**

There must be undefined terms in any branch of mathematics, as Aristotle had stressed, and **all that one can require of these lines is that they satisfy the axioms.**"

[2, page 217] "As we have explained, totally **different interpretations are possible because undefined terms** are necessarily present in any axiomatic development. These interpretations are called models. Thus, what we havc seen is that a branch of mathematics created with one physical meaning intended may apply to and entirely different physical or mathematical situation."

[2, page 229] "**The first concerns the necessity for undefined terms. Since mathematics is independent of other subjects, a definition must be in terms of other mathematical concepts. Such a process would lead to an infinite regress of definitions. The resolution of this difficulty is that the basic concepts must be undefined.** How then can they be used? How does one know what facts about them can be asserted? The answer is that the axioms make assertions about the undefined (and defined) concepts and so the axioms tell us what can be asserted. Thus, if point and line are undefined, the axiom that two points determine a unique line and the axiom that three points determine a plane furnish assertions that can be used to deduce further results about point, line and plane. Though Aristotle in the *Organon*, Pascal in *Treatise on the Geometrical Spirit*, and Leibniz in *Monadology* had emphasized the need for undefined terms, mathematicians peculiarly overlooked this fact and consequently gave definitions that were meaningless. Joseph-Diaz Gergonne (1771-1859) in the early nineteenth century pointed out that the axioms tell us what we may assert about the undefined terms; they give what may be called an implicit definition. It was not until Moritz Pasch reaffirmed the need for undefined terms in 1882 that mathematicians took this matter seriously.

The fact that any deductive system must contain undefined terms that can be interpreted to be anything that satisfies the axioms, introduced into mathematics a new level of abstraction. This was recognized rather early by Hermann Grassman in his *Theory of Linear Extension* (1844), who pointed out that **geometry must be distinguished from the study of physical space. Geometry is a purely mathematical structure which can apply to physical space but is not limited to that interpretations."**

[2, pages 230, 231] "We have already noted that the creation of non-Euclidean geometry had forced the realization that mathematics is man-made and describes only approximately what happens in the real world. The description is remarkably successful, but it is not truth in the sense of representing the inherent structure of the universe and therefore not necessarily consistent. **Indeed, the axiomatic movement of the late nineteenth century made mathematicians realize the gulf that separated mathematics from the real world. Every axiom system contains undefined terms whose properties are specified only by the axioms.** The meaning of these terms is not fixed, even though intuitively we have numbers or points or lines in mind. To be sure, the axioms are supposed to fix properties so that these terms do indeed possess the properties that we intuitively associate with them."

From those citations we can see followed points:

1) frequent usage of term "undefined" for notion, concept, idea, term;
2) different description meaning of this kind of concepts:

a) with reference to intuition;
b) with reference to everyday experience;
c) defined by followed axiom, undefined notion is not fix, can be interpreted by different physical situation with requirement to satisfy axiom only;

3) connection undefined notion within other notions, without consideration correspondence to physical object or phenomena;
4) underlined independency notions from reality;

The better definition of "undefined notion, concept, term, idea" is "initial notion," and any notion should be considered on inconsistency. Let's consider notion on infinity as initial notion and analyze in detail its inconsistency. Later, in Section 7, we will discuss the initial notions of Euclid's geometry.

In [10] Eugene Wigner defines mathematics as " ...mathematics is the science of skillful operation of concepts and rules invented just for this purpose. The principal emphasis is on the invention of concepts." Poincare defines mathematics as the science, which gives the same name to different things. It is usual to find the strength of mathematics in unification, generalization, abstraction, and idealization, but here we can lose the quality of an object. It is well known that each physical notion has two aspects: quantity and quality, quantity is expressed numerically, and quality is expressed by dimensionality. To operate on physical notions, we should know theirs quality, for example, we can add physical quantities having the same dimension. But this is not enough: we can write the sum of densities, but we do not have any physical process to double density, consequently, this sum does not make sense. We can multiply values different quality, for example, mass and speed, and get value third kind of quality, in this case – impulse. Some mathematical structures are defined as consisting of one kind of physical quantity. Group theory defines a binary operation on just one kind of object for both operands, and the result of the operation is of the same type of physical quantity. From the physical point of view, physical groups are a very special case. The well known physical redefinition is: $h = c = 1$, along with mention of an imaginary system of reference. This is absurd from the points of view of logic, mathematics, and physics, namely, the equivalence of very small value h having one type of dimensionality, and the very big value c with having a different dimensionality, and the dimensionless unit. We have to emphasize importance of dimensionality: it defines quality of physical quantity and binds it with reality. If mathematics works with dimensionless numbers, then it is up to physics to validate the meaning of its equations by using the dimensions of its physical values.

50 years after Eugene Wigner's work, in "On the Reasonable and Unreasonable Effectiveness of Mathematics in Classical and Quantum Physics", (The title of this article is a paraphrase of the title of Eugene Wigner's famous paper, "The Unreasonable Effectiveness of Mathematics in the Natural Sciences,") Arkady Plotnitsky writes [Foundations of Physics Journal, March 2011, Volume 41, Issue 3, pp 466-491]:

"Quantum mechanics, thus, and then higher-level quantum theories continue classical physics insofar as it is, just as classical physics, from Galileo on, and then relativity has been, the experimental-mathematical science of nature. However, quantum theory, at least, again, in the interpretations of the type discussed here, breaks with both classical physics and relativity by establishing radically new relationships between mathematics and physics, or mathematics and nature. The mathematics of quantum theory is able to predict correctly the experimental data in question without offering and even preventing the description of the physical processes responsible for these data."

We see that after 50 years, it was changed nothing – quantum mechanics is "the experimental-mathematical science of nature," no problems, everything is logical.

Now we consider very important overlooked aspect of the mathematical object: dimensionless numbers. Dimensionless numbers have qualities (properties) too. We have the whole numbers: 1, 2, 3, 4, 5, 6, 7, 8, 9 – call these, if you wish, Mathematics' atomic table – they are unique, each of them equal to itself and between different ones there exists rigorous inequality, and their different appearances correspond to different meanings. The last of the properties is one of strings so that $1 > 0.999 \ldots$, because right and left sides this inequality has different appearance. If we would accepted equality $1 = 0.999 \ldots$, then we will accept illogicality that on infinity $9 = 10$, again – part equal to whole. The rest of all mathematics consists of expressions – bigger numbers then 9 are expressions, and the next quality of numbers – rational – are expressions, with uniqueness of expressions guaranteed by the uniqueness of the nine digits. It is easy to see that mathematics is the science of the manipulation of symbolic expressions.

The next numbers with a different quality are negative numbers, which for the first time were utilized in India to calculate of a money debt. It is clear that quality "negative" is a man-made for man concept, which does not exists in reality. We do not have positive and negative charges; we have one-kind of and second kind of charges, and use negative numbers to "automate" calculation of the direction of the interaction of charges. The generalization of the sum of several the same operands brings to us multiplication. From this, multiplication by unity and negative numbers does not make sense. The next level of abstract generalization is the postulation of multiplication as a second kind of operation, simply different from summation. After that we get the postulation of multiplication by unity and negative numbers. We have to note, that this increasing of the level of abstraction leading to new notions of quality, and the mixing of the notions of different level of abstraction involve us in implicit contradictions even in the abstract world of mathematics and hide the inconsistency problems in physics.

The next level of abstraction is the root operation, and the next - complex numbers, but this amounts to "pipe dreams" as Roger Penrose points out [11].

The zero plays a special role, as a digit, it is used to calculate empty space in mathematical expressions, to hide opposite objects, and to give birth to not-zero structures and physical objects.

Geometry plays a particular role in mathematics and physics. Geometry is an abstract science with objects obtained from the abstraction and idealization of properties of solid bodies. Geometry has given birth to a lot of different kinds of abstract objects and made simple use of the following objects: irrational, rational, complex numbers and so on. The unification of the geometry of a solid body on space yields us more problems than advantages. The main problem is: space does not interact with any body or substance; this is the main property of space. Ivchenkov has shown [12] that Eddington's observations were within measurement error bounds. We still do not have any physical observations of any interaction with space. Geometry has within itself an inconsistency; namely, there is a point as an object without parts, as a geometrical zero.

All of the above show the importance of the investigation of the consistency of initial notions and concepts. The next step is the redefinition of all old notions to obtain confidence in the consistency of the new notions as in the cases of the old ones. Let's consider the redefinition of physical notions.

One of forms of the first postulate of special relativity theory (SRT) – namely, independency velocity of light from any inertial system of reference, that is, the express redefinition of relative value of the velocity of light, namely, light velocity as a single and absolute value. This looks like "redefinition without definition," because SRT does not define notion of "absolute velocity." The definition of 4-velocity gives us a unit as four-velocity by definition, and the 4-acceleration which always orthogonal to 4-velocity by definition [13].

The well known Heisenberg uncertainty principle states that certain pairs of physical values, such as energy and time, and coordinate and impulse, cannot be simultaneously measured with arbitrarily high precision [14]. This principle has an inconsistency, because impulse is function of the coordinates, but the energy is function of time. If in this principle, there were involved different quantities like a quasi-impulse which was independent of the coordinates, and then we could have at least two incommensurable notions of impulse, and two incommensurable notions of energy.

The very interesting generalization-redefinition of an arbitrary translation of vector as parallel motion in non-Euclidian geometry is one we can investigate in [11, 13]. It is convenient to follow Penrose [11], where we can see diagrams of the surface geometry of a sphere. In spherical geometry, the analogue of the straight line in Euclidian geometry is the sphere's surface meridian. These two are very different objects: the radius of curvature of meridian is finite, but radius of curvature of straight line is infinite. They are similar because they express the shortest distance between two points, one in Euclidean two-space and the other in the two-space of the sphere's surface. Parallel motion in Euclidian geometry preserves the direction of a vector, and when a vector is translated along a close path, the vector will coincide with itself. According to Penrose and others, in non-Euclidian geometry, parallel translation of a tangent vector along a close path need not bring superposition of vector with itself. Let's consider some objections.

First of all, Penrose draws a tangent vector in 3-dimensional space, and maintains that on the sphere surface lays the tail of the vector. But the tangent vector of spherical surface must belong to the sphere's parts, and it is part of meridian, because meridian is a straight line of its sphere. Secondly, Penrose did not define equivalence of direction on sphere. Well known, that all meridians on sphere intersect each other in two points. This means that we can keep track of the direction of the tangent vector in attempting the parallel translation, along one meridian only. In this case, we get the superposition of the vector on itself. If we try to move a vector out of the starting meridian, we will have to move the vector to another meridian, which intersects first one, and this vector will not coincide with itself at the point of intersection of the two meridians, because we have changed its direction. In Euclidian geometry we get similar result if will change the direction of vector, and this translation is not parallel. On this arbitrary-parallel translation was built tensor analysis.

Thus, we have at least two problems: the consistency of initial notions, and the consistency of redefined notions. We can see hierarchical relations between notions, and mixing notions of different level leads to inconsistency.

2. World Well Known Notion of Infinity.

Notion of infinity one of the oldest notion which usage is very broad: in mythology, religions, arts, science, especially, in mathematics and physics, and from Anaximander (610 – 546 BC) and Aristotle (384 BCE – 322 BCE) to Sergeyev (1963-present), author of the newest theory of infinity. There are a lot books which dedicated to infinity [15-20] and more. All those books are about a big men's love to paradoxes and completely ignoring of logics – ignoring the obvious inconsistency of notion of infinity.

Infinity is not exist in reality, and exists in man's imagination only. We cannot operate with infinity in logical reasoning, but when men try to do that it will lead to inconsistencies. A lot attempting to operate notion of infinity looks like operation with big number – not number is taken as number – it is reason initial inconsistencies. There were invented a lot symbol of infinity (Section 3), and algebra infinity (Section 4) has peculiar properties and was mixed with algebra of number. The special "logic" of infinity has one "argument" – "because it is infinity," and this is argument to justify illogicality. Nobody, nowhere, and never can take a look or do something "on infinity," even one-to-one correspondence. We can not make decision about any properties "on infinity" – like "all numbers" and "one-to-one correspondence." We can believe it only, but it is religion activity, not scientific activity. If we do that then we have to pay too much: abandon logics and take illogic.

Notion of infinity cannot be corresponding to natural object, and was created by language parts: prefix and suffices:

Finite – infinite, end – endless, bounded – unbounded, boundless, limit – limitless, unlimited and so on.

Understanding this fact by Aristotle was expressed in notions of absolute and potential infinity. In the Sergeyev's newest theory of infinity we see in first postulate impossibility infinite action by men and machine.

Notion of infinity was used to identify contradictory concepts like infinitesimal, and spatial, time infinity, continuity, contradictory expression like "infinite number" : infinite is not number, and number is not infinite.

There are different signs of infinite and they algebras which mixed infinite and numbers, and create illusion of applicability of infinity.

Usage argument as "because infinite" – to argue different "logic" of conclusion: part equal whole, infinite means "all" and so on.

In [16, pages viiii-ix] Eli Maor writes:

"But however we look at the infinite, we are ultimately led back to mathematics, for it is here that the concept of infinity has its deepest roots. According to one view, mathematics is the science of infinity. In the Encyclopedic Dictionary of Mathematics, a compendium recently compiled by the Mathematical Society of Japan (English translation published by The MIT Press, 1980), the words "infinity," "infinite," and "infinitesimal" appear no fewer than 50 times in the index. Indeed, it is hard to see how mathematics could exist without the notion of infinity …." We can note that in second edition mention dictionary, in 1995, in index are 67 those items [21].

In [18, page 5] David Foster writes:

" There is such thing as an historian of mathematics. Here is a nice opening-type quotation from one such historian in the 1930s:

> One conclusion appears to be inescapable: without a consistent theory of the mathematical infinite there is no theory of irrationals; without a theory of irrationals there is no mathematical analysis in any form even remotely resembling what we now have; and finally, without analysis the major part of mathematics – including geometry and most of applied mathematics – as it now exists would cease to exist. The most important task confronting mathematicians would therefore seem to be the construction of a satisfactory theory of the infinite. Cantor attempted this, with what success will be seen later."

It is extreme opinion, because we cannot remove infinity and derived branch of mathematics from history. We can see inconsistency of notion of infinity and proximity of math model with usage contradictory concepts.

Frederick Engels, in Anti-Dühring [22], 1877, (Part I: Philosophy, of V. Philosophy Nature. Time and Space) writes:

"Infinity is a contradiction, and is full of contradictions. From the outset it is a contradiction that an infinity is composed of nothing but finites, and yet this is the case. The limitedness of the material world leads no less to contradictions than its unlimitedness, and every attempt to get over these contradictions leads, as we have seen, to new and worse contradictions. It is just *because* infinity is a contradiction that it is an infinite process, unrolling endlessly in time and in space. The removal of the contradiction would be the end of infinity. Hegel saw this quite correctly, and for that reason treated with well-merited contempt the gentlemen who subtilised over this contradiction."

We can note from this citation well known statements:

1) explicitly call infinity as contradiction;
2) underlined that infinity composed from finites which have contradiction too;
3) explanation of infinity as process in time and space; it is base of intuition of infinity: if we can do from n to n+1, and from m to m/2 then we can do that "to infinity."
4) the most important conclusion is that "The removal of the contradiction would be the end of infinity.."

The end of infinity should be provided from logic point of view, in logical science or science with pretension on logicality. In art, poetry, religion infinity can be used without any exclusion.

To show necessity excluding infinity from logic, we need explicitly define infinite's contradictions, namely:

1) The terms infinite, endless, unlimitedness and so on was created by language's tool: by prefix and suffix, and cannot be correspond to reality, infinity is correspond to our imagination or faith ONLY.
2) Man cannot operate with infinity as with infinite object or with infinite number of operation, but can operate with term or symbols of infinity wherein he completely ignore difference between symbols manipulation and action with consistence this symbols.
3) Algebra does not work for infinity: infinity and any its symbol is not number, infinity does not have quantitative properties and can not be used with mathematical operations.

Now we can repeat that "every attempt to get over these contradictions leads, as we have seen, to new and worse contradictions." It we will see in Sergeyev's theory of Grossone later.

Development of infinity brings different "quality" of infinity: actual, potential, ordinality, cardinality, transfinite, countable, uncountable, and so on, and last of one – Sergeyev's numerals – Grossone.

Especially, notion of infinity was developed in the Cantor's theory of infinite set.

"Anti-Dühring" by Frederick Engels, was written in 1877, around this time 1873-1884, George Cantor had created his set theory.

The importance of those notions cannot be overestimated, because we can find them in almost all mathematical and physical theory, and from the discrepancies of infinity, the infinitesimal, and the zero, there result inconsistencies in these theories. In short, we can refer to the contradiction of infinity by the term "finite infinity," the contradiction of the infinitesimal as "the notion which is equal and not equal of zero at the same time," and the contradiction of zero as "the declaration of the existence of the non-existent."

Mathematicians have noticed long ago the contradiction in the expression – "infinite number," because any number is a finite object, and consequently, infinity is not a number. The infinite sequence of steps and the infinitesimal can be seen in paradox due to Zeno of Elea, of Achilles and the tortoise, Dichotomy (490 BC–430 BC). If we will consider a finite number of steps of Achilles and tortoise, having finite distances between them, then Achilles will catch the tortoise in finite number of steps. But if we consider the division of this finite distance into an infinite number of parts, then we arrive at a contradiction: the steps become infinitesimal, but not zero, because an infinite sum of zeros equal zero, but an infinite sum of non-zero constant length steps equals an infinite distance, or we would accepted equivalence of part to whole in case of steps with variable length.

Let us consider the well known explicit redefinition of infinity by Cantor: he defines the number of elements of infinite set as omega or aleph-null, and operates with them as numbers. This generalization – redefinition of a non-number as number by Cantor, Hilbert refers to as a "mathematical paradise."

Solutions of the Dichotomy paradox often are expressed as a limit of infinite sum of inverse powers of two. If we accept this, then we accept an illogical result: on infinity this sum, half would equal to the whole. This contradiction we find in limit theory and mathematical analysis as the equivalence of the part to the whole. In set theory, the equivalence of a part of a set to the whole is used to give the definition of an infinite set [23]. It is completely illogical: from this we get the consequences – the part is more than itself, and the whole is less itself, and so we have lost the equivalency itself for the part and the whole (part equal to part, and whole equal to whole).

The notion of zero has three main meanings: 1) the physical – nothing, empty space, not existent something; 2) the geometrical – dimensionless point, which does not have any parts according to Euclid's Elements; 3) mathematical – empty space in expression, digit, number. The physical meaning of zero conflicts with geometrical and mathematical ones, where objects of zero size claim to be existent objects. We can see in one the algebra's axiom three contradictions: a) declaration of the existence of an element zero; b) we can add zero to another number – summing using empty space; c) in the binary operator – addition, we can use with one operand, because zero is empty space. One objection is that after the postulation of existence of the zero-element, cases b) and c) became valid. We can point that zero is involved in the manipulation of expressions, and exactly for the "automatic" calculation of empty space characterizations. For this goal of calculation, they were defined and this concludes our discussion concerning operations involving zero.

3. Set Theory.

George Cantor had created his set theory in 1873-1884. He defines an infinite set as set which is equal its subset by one-to-one correspondence. This Cantor's set doctrine was called as naive, because was found Russell's paradox. In improved set theory on ZF-axioms was involved axiom of existence of infinite set. One of the main concepts was one-to-one correspondence (1-1-C) and infinite expression was marked by three dots … Also, very important axiom was "intuitively" accepted: "infinite" means "all." This "axiom" contradicts to one-to-one correspondence. Indeed, let's take a look on very well known expression of 1-1-C between natural number and even number:

1 2 3 4 5 6 7 8 9 10 … n …

2 4 6 8 10 12 14 16 18 20 … 2n …

Here, because 1-1-C and "infinite-all" axiom, we have definition of infinite set as set which equal its subset: infinite series of natural numbers equal infinite series of even numbers. In "all" finite cases we can note that this two series have equal subseries only: here, for first 10 number of finite series, there is subseries 2 4 6 8 10, and it is half of each of them. If we follow 1-1-C then each even number twice more than its natural number in series, in any place in series. For "infinite-all" case it is not work: if it is "all" then it is not 1-1-C, and if it is 1-1-C we cannot achieve "all." It is very frequently used expression: what is work in finite case, does not work for infinity – it is conclusion from "infinite-all" axiom.

First of all, why we have to accept "infinite-all" axiom? Nobody, nowhere, never can check this "on infinity." Secondly, in finite case 1-1-C does not give us equivalency of series of natural number and series its even number: why we should accept equivalence on infinite? Because it is infinity and on infinity we can get what we wish (case A for 2 + 2 = ? What do you wish?). Expression "all number" does not make sense in our finite world.

Now we can look at next 1-1-C expression with "infinite-all" axiom:

1 2 3 4 5 6 7 8 9 10 …
1 2 3 4 5 6 7 8 9 10 …

From this we can conclude that the whole natural number is equal itself. It means that reference to infinity give us two different conclusions:

Part is equal whole (1-1-C for n … 2n) and
Whole is equal itself;

In the mathematical set theory is accepted first one, we can accept second one because everyday experience, but, from logic point of view, we have to deny reference to infinity and deny both of them for infinite expressions. Thus any infinite expression has inconsistency, and can not be used for logical reasoning.

We can find a lot example of mathematical "equivalence" of part and whole:

Similar triangles, projective geometry (see Section 7), and so on.

Cantor's doctrine is not theory – because based on the faith in infinite object and possibility to operate by infinite object, on the faith that part can be equal to whole.

We have to note, that in mathematics it was very often contrasted an obvious clear Descartes' mathematical intuition to the man's senses which do not have other goal but make mistakes. It is true except one notion, related to the man's senses – notion of continuity. We see and feel by hand continuous surface of solid body, we see continuous surface of liquid. Continuity plays prima role and the math intuition tries to confirm this notion. As result of an intuition work we have notion of infinitesimal value, and implicit or explicit acceptance of an infinite division of finite math object: line's segment, geometrical figures and so on. As infinite, infinitesimal value has inconsistency too: in the same time value of infinitesimal object equal and not equal zero. Reference to infinite division of finite distance we see in Zeno's paradoxes. It is reason of paradoxes, and resolution of paradox is denying the infinite division of finite distance. Resolution of paradox by ignoring this inconsistency: involved notion of velocity, reference to the limit's theory with infinite expression, - is substitution of an one illogicality by another inconsistency.

4. Signs of Infinity.

Symbols of infinity are:

Uroboros — snake eating its own tail (see title page).

∞	infinity used in mathematics, not set theory.
c	is the continuum $c = 2^{\aleph_0} = \aleph_1$.
ω	infinite ordinals, (or ω0,) has cardinality $\aleph_0$.
$\aleph_0$	the smallest infinite cardinal number, aleph-zero.
$\aleph_1$	is the cardinality of the set of all countable ordinal numbers, called $\boldsymbol{\omega}_1$ or (sometimes) **Ω**.
①	Sergeyev's Grossone with different algebra. [24].

Those are not all symbols of infinity, but main of them which are used more often than others.

We have to notice that we can not operate any kind of infinity, but we can operate by symbols, and establishment of the infinity's symbols create illusion possibility of operation by infinity. Algebra of the symbolic infinity will consider in next section.

From historical point of view, the invention of writing to mankind was huge step in its progress. Applying of symbols in all fields of men's activities makes the usual an accordance of words, languages, theories to reality. Long time of applying words lead to illusion that notion created by formal way could be corresponding to real object. Existing invisible reality makes easier acceptance abstract notion as real object. It can bc illustrate in steps: real object – word – derived notion – new notion – invisible expected object – new not existed "real" object. Infinity as word or notion, and all its symbols and abbreviations do not have a corresponded real object.

5. Algebra of Infinity.

Algebra of infinity we can find in Encyclopedic Dictionary of Mathematics [21, page 326]:

$a * (\pm \infty) = \pm \infty, a > 0;$

$a * (+ \infty) = - \infty, a < 0;$

$a * (- \infty) = + \infty, a < 0;$

$a \pm \infty = \pm \infty;$

$a / (\pm \infty) = 0;$

The cases

$0 * (\pm \infty), + \infty + (- \infty), (\pm \infty) / \pm \infty$ excluded.

Excluded case $0 * (\pm \infty)$ is questioning, because this multiplication can be substitute by infinite sum:

$$\sum_{1}^{\infty} 0 = 0;$$

This sum obvious equal zero, but not for everybody. Physicist Dirac invented Dirac's delta-function, which equal infinity in one point, and equal zero in all other points of numerical axis. Integral in infinite range $\pm\infty$ from this function equal unit 1. We can note that if dx more than zero, then integral should be equal infinity. Width one point, where delta function equal infinite, is equal zero, or we can suppose that $\infty * dx = \infty * 0 = 1$ per Dirac's proposition. So, we have three variants: A) exclusion of $0 * (\pm\infty)$, B) $0 * (\pm\infty) = 0$, and C) $0 * (\pm\infty) = 1$. It is very powerful algebra for question: What are you wish?

Dirac's invention was developed by mathematicians in whole branch of mathematics: generalized functions. Strictly speaking, manipulation with notion infinity in this case was used to get desire result, because on infinity everything possible and nobody cannot check on infinity.

It is easy to see that this algebra different from regular algebra, and obvious reason is because we use infinity. It is a mixing of numbers and not number objects. In set theory difference infinity from number underline by usage notion of power of an infinite set. Mixing incomparable notions bring inconsistency to this algebra.

We have to notice and problem of regular algebra. Core of algebra is using symbolic expression and manipulate and transform them from one to other. In regular algebra we use symbol of zero 0: $a + 0 = a$. Zero is symbol empty place, and if we define binary operation +, then we have to use two operand, not empty space: $a + \ = a$. There are more illogical expressions in regular algebra: $a*1 = a$; $a/1 = a$; multiplication and division on unite does not make sense, but manipulation in this way does not change numerical value of expression and it is reason why we can accept them. But formal usage zero in expression does not give us base to declare zero as number. If we do so – call empty space as number, then we bring inconsistency to algebra, and physics can wait for existence of a substantial vacuum.

Sergeyev's algebra per [24] is:

$$0 \cdot ① = ① \cdot 0 = 0,\ ① - ① = 0,\ \frac{①}{①} = 1,\ ①^0 = 1,\ 1^{①} = 1,\ 0^{①} = 0,$$

$$0 \cdot ①^{-1} = ①^{-1} \cdot 0 = 0,\ ①^{-1} > 0,\ ①^{-2} > 0,\ ①^{-1} - ①^{-1} = 0,$$

$$\frac{①^{-1}}{①^{-1}} = 1,\ \frac{①^{-2}}{①^{-2}} = 1,\ \left(①^{-1}\right)^0 = 1,\ ① \cdot ①^{-1} = 1,\ ① \cdot ①^{-2} = ①^{-1}.$$

We can see from this algebra that it is usual algebra of number despite Sergeyev's declaration that its Grossone is numeral of infinity. The Sergeyev's theory of Grossone will be considered inconsistencies in next section.

6. Sergeyev's Grossone Theory.

The Sergeyev's Grossone theory [24-27] is the newest theory of infinity. Author sincerely believes in his theory: he pays money for three patents on infinite computer, in countries USA, Italy and Russia [The Infinity Computer (European patent EP 1728149, Russian patent 2395111, US patent 7,860,914)[25]. Sergeyev invented new symbol – Grossone ①**,** create algebra with this symbol, declare that part is less than whole, and, in the same time, declare that his theory is not contradict to Cantor's theory, where infinite set defined as equality subset to whole set per one-to-one correspondence.

Let's consider the Grossone theory in detail beginning from its postulates. Postulates declare [26]:

"**Postulate 1.** *Existence of infinite and infinitesimal objects is postulated but it is also accepted that human beings and machines are able to execute only a finite number of operations.*
Postulate 2. *It is not discussed* ***what are*** *the mathematical objects we deal with; we just construct more powerful tools that allow us to improve our capacities to observe and to describe properties of mathematical objects.*
Postulate 3. *The principle formulated by Ancient Greeks 'The part is less than the whole' is applied to all numbers (finite, infinite, and infinitesimal) and to all sets and processes (finite and infinite).*

Due to this declared applied statement, such traditional concepts as bijection, numerable and continuum sets, cardinal and ordinal numbers are not applied when one works with the Infinity Computer because they belong to Cantor's approach having significantly more theoretical character and based on different assumptions. However, the methodology used by the Infinity Computer does not contradict Cantor. In contrast, it evolves his deep ideas regarding existence of different infinite numbers in a more practical way."

First postulate declares existence "infinite and infinitesimal objects," but what is it, those objects? In third postulate those objects mentioned in list of "all numbers," included finite number, and "all sets and processes." Second statement of first postulate "accepted" only finite number of operation for men and machines. Second postulate refuse even to discuss "what are math objects," and "permit" "just construct more powerful tools" to observe and describe properties of mathematical objects. Indeed, very "powerful" position to observe and describe properties of prohibited objects. It is next "level of abstraction" from abstraction: properties exist but objects are not, even those existences were declared by first postulate. As predict Engels, accumulation more inconsistencies begin from the postulates of the new theory of infinity. Third postulate declares that "The part is less than the whole" and that is contradict to Cantor's set theory.

There are even more, let's take a look. In [27, pages 27, 28] Sergeyev writes:

"Let us remind one more famous example related to the one-to-one correspondence and taking its origins in studies of Galileo Galilei: even numbers can be put in a one-to-one correspondence with all natural numbers in spite of the fact that they are a part of them:

$$\begin{array}{lccccccccc} \textit{even numbers:} & 2, & 4, & 6, & 8, & 10, & 12, & \dots & & \\ & \updownarrow & \updownarrow & \updownarrow & \updownarrow & \updownarrow & \updownarrow & & & (36) \\ \textit{natural numbers} & 1, & 2, & 3, & 4, & 5, & 6, & \dots & & \end{array}$$

Again, our view on this situation is different **since we cannot establish a one-to-one correspondence between the sets because they are infinite and we, due to Postulate 1, are able to execute only a finite number of operations. We cannot use the one-to-one correspondence as an executable operation when it is necessary to work with infinite sets."**

Bold text is our underline to pay attention. We continue citation:

"However, we already know that the number of elements of the set of natural numbers is equal to ① and ① is even. Since the number of elements of the set of even numbers is equal to $\frac{①}{2}$, we can write down not only initial (as it is usually done traditionally) but also the final part of (36)

$$\begin{array}{cccccccccccc} 2, & 4, & 6, & 8, & 10, & 12, & \dots & ①\text{-}4, & ①\text{-}2, & ① & \\ \updownarrow & \updownarrow & \updownarrow & \updownarrow & \updownarrow & \updownarrow & & \updownarrow & \updownarrow & \updownarrow & (37) \\ 1, & 2, & 3, & 4, & 5, & 6, & \dots & \frac{①}{2}\text{-}2 & \frac{①}{2}\text{-}1 & \frac{①}{2} & \end{array}$$

concluding so (36) in a complete accordance with Postulate 3. Note that record (37) does not affirms that we have established the one-to-one correspondence among all even numbers and a half of natural ones. We cannot do this due to Postulate 1. The symbols ' … ' indicate an infinite number of numbers and we can execute only a finite number of operations. However, record (37) affirms that for any even number expressible in the chosen numeral system it is possible to indicate the corresponding natural number in the lower row of (37)."

It easy to see, per Sergeyev's theory we cannot do something per Postulate 1, but the same we can do per Postulate 3. It looks like the Sergeyev's Grossone is sometimes number, but sometimes it is infinity. Thus, the Sergeyev's theory of Grossone brings more inconsistencies to concept of infinity which was inconsistent and before new theory. The Engels' forecast in 1877 about removing of the contradiction of infinite was realized in 21-st century.

7. Usage of infinite expressions: 1 ? 0.999 …

We will consider in this section relation between unit 1 and infinite expression 0.999 … Consideration will base on article 0.999 … from Wikipedia [28]. As it was done in article, we use reference to infinity, and infinite expressions. It will show that 1 > 0.999 … in contrast to article where 1 = 0.999 … We propose conclusion that reference to infinity and usage of the infinite expression should be excluded from logical reasoning, because usage of reference on infinity and the infinite expression lead to controversial conclusion.

7.1 Algebraic proofs

Here and below per [28] "Algebraic proofs showing that 0.999 … represents the number 1 use concepts such as fractions, long division, and digit manipulation to build transformations preserving equality from 0.999 … to 1."

Fractions and long division

"Onc reason that infinite decimals are a necessary extension of finite decimals is to represent fractions. Using long division, a simple division of integers like $^1/_9$ becomes a recurring decimal, 0.111 …, in which the digits repeat without end. This decimal yields a quick proof for 0.999 … = 1. Multiplication of 9 times 1 produces 9 in each digit, so 9 × 0.111 … equals 0.999 … and 9 × $^1/_9$ equals 1, so 0.999 … = 1:

$$\frac{1}{9} = 0.111...$$

$$9 \times \frac{1}{9} = 9 \times 0.111...$$

$$1 = 0.999...$$

Another form of this proof multiplies $^1/_3$ = 0.333 … by 3."

7.2. Discussion of "Fractions and long division."

Statement above includes logical error: usage unproven expression:

$$\frac{1}{9} = 0.111... \text{ and } \frac{1}{3} = 0.333...;$$

We can write:

$$\frac{1}{9} = 0.1 + \frac{1}{90} \; and \; \frac{1}{3} = 0.3 + \frac{1}{30};$$

$$\frac{1}{9} = 0.11 + \frac{1}{900} \textit{ and } \frac{1}{3} = 0.33 + \frac{1}{300};$$

And so on to infinity:

$$\frac{1}{9} = 0.111... + \frac{1}{900...} \textit{ and } \frac{1}{3} = 0.333... + \frac{1}{300....};$$

Let's consider inequalities:

$$300 < 900 < \infty;$$

$$\frac{1}{300} > \frac{1}{900} > \frac{1}{\infty} = 0;$$

$$\frac{1}{300 \cdot 10} > \frac{1}{900 \cdot 10} > \frac{1}{\infty \cdot 10} = \frac{1}{\infty} = 0;$$

$$\frac{1}{300...} > \frac{1}{900...} > \frac{1}{\infty} = 0;$$

if so then equalities:

$$\frac{1}{9} = 0.111... + \frac{1}{900...} \textit{ and } \frac{1}{3} = 0.333... + \frac{1}{300....};$$

Become inequalities:

$$\frac{1}{9} > 0.111... \textit{ and } \frac{1}{3} > 0.333...;$$

Because $\frac{1}{300...}$ *and* $\frac{1}{900...}$ more than zero.

And: $1 > 0.999\ldots$ from both inequalities.

7.3 Digit manipulation

[28] "When a number in decimal notation is multiplied by 10, the digits do not change but each digit moves one place to the left. Thus 10 × 0.999 … equals 9.999 …, which is 9 greater than the original number. To see this, consider that in subtracting 0.999 … from 9.999 …, each of the digits after the decimal separator cancels, i.e. the result is 9 − 9 = 0 for each such digit. The final step uses algebra:

$x = 0.999\ldots$

$10x = 9.999\ldots$

$10x - x = 9.999\ldots - 0.999\ldots$

$9x = 9$

$x = 1$ ".

7.4 Discussion of Digit manipulation.

There is used arithmetic's operation with infinite expression:

10*0.999 … and 9.999 … - 0.999 …

Can we operate with infinite expressions? Let's start do this and divide two equalities.

$$\frac{10x = 9.999\ldots}{x = 0.999\ldots};$$

$$10 = \frac{9.999\ldots}{0.999\ldots} = \frac{999\ldots}{999\ldots} = 1;$$

We get 10 = 1 by using infinite expressions.

7.5 Discussion

[28]"Although these proofs demonstrate that 0.999 … = 1, the extent to which they explain the equation depends on the audience. In introductory arithmetic, such proofs help explain why 0.999 … = 1 but 0.333 … < 0.34. And in introductory algebra, the proofs help explain why the general method of converting between fractions and repeating decimals works. But the proofs shed little light on the fundamental relationship between decimals and the numbers they represent, which underlies the question of how two different decimals can be said to be equal at all.

Once a representation scheme is defined, it can be used to justify the rules of decimal arithmetic used in the above proofs. Moreover, one can directly demonstrate that the decimals 0.999 … and 1.000 … both represent the same real number; it is built into the definition. This is done below."

7.6 Discussion of Discussion.

We can note that if 0.999 … and 1.000 … are the same real number by definition, it is problem and for real number too. Let's consider motivation this problem – equality 1.000 and 0.999 …

We have nine signs or numerals or digit for number (for convenient):

I	1	digit, numeral, number one
II	2	digit, numeral, number two
III	3	digit, numeral, number three
IIII	4	digit, numeral, number four
IIIII	5	digit, numeral, number five
IIIIII	6	digit, numeral, number six
IIIIIII	7	digit, numeral, number seven
IIIIIIII	8	digit, numeral, number eight
IIIIIIIII	9	digit, numeral, number nine

Other number is expression like 100 = (9 + 1)*(9 + 1). Zero 0 is not number; it is sign of empty space in positional notation. Declaration of zero as number is one more illogicality in math. Usage those nine signs in expression give us unique expression. Different appearance expression with those signs should give us different value, or 0.999 … and 1.000 … should be different by appearance. This issue will be discussed later.

7.7 Analytic proofs

[28] "Since the question of 0.999 … does not affect the formal development of mathematics, it can be postponed until one proves the standard theorems of real analysis. One requirement is to characterize real numbers that can be written in decimal notation, consisting of an optional sign, a finite sequence of any number of digits forming an integer part, a decimal separator, and a sequence of digits forming a fractional part. For the purpose of discussing 0.999 …, the integer part can be summarized as b_0 and one can neglect negatives, so a decimal expansion has the form

$b_0.b_1\, b_2\, b_3\, b_4\, b_5 \ldots$

It should be noted that the fraction part, unlike the integer part, is not limited to a finite number of digits. This is a positional notation, so for example the digit 5 in 500 contributes ten times as much as the 5 in 50, and the 5 in 0.05 contributes one tenth as much as the 5 in 0.5."

Notes: The question of 0.999 … DOES AFFECT the formal development of mathematics because it is infinite expression and usage an infinite expression give us inconsistency as was shown above in 7.2 – in 7.1. was proved that 0.999 … =1, and in 7.2. was proved that 1 > 0.999 …

7.8. Infinite series and sequences.

[28]"The statement that 0.999 … = 1 can itself be interpreted and proven as a limit:

$$0.999... = \lim_{n\to\infty} 0.99...9_n = \lim_{n\to\infty}\sum_{k=1}^{n}\frac{9}{10^k} = \lim_{n\to\infty}\left(1-\frac{1}{10^n}\right) = 1-\lim_{n\to\infty}\frac{1}{10^n} = 1;$$

The last step, that $\frac{1}{10^n} \to 0 \quad as \quad n \to \infty$, is often justified by the Archimedean property of the real numbers."

We can note that $\frac{1}{10^n} > 0 \; when \; n \to \infty$ in similar way as in 7.2.

$$\frac{1}{10} > \frac{1}{\infty} = 0;$$

$$\frac{1}{10\cdot 10} > \frac{1}{\infty\cdot 10} = \frac{1}{\infty} = 0;$$

$$\frac{1}{10\cdot 10\cdot 10} > \frac{1}{\infty\cdot 10\cdot 10} = \frac{1}{\infty} = 0;$$

…

$$\frac{1}{10...} > \frac{1}{\infty} = 0.$$

7.9. Dedekind cuts.

Geometric presentation of the Dedekind's cut uses the inconsistent geometric notion of "point"- object without parts, the symbolic presentation uses finite alphabet and even in infinite expression of real number can cover countable set of number only, between those numbers there are infinite points from uncountable set of points. And more, we can not distinguish even countable infinite set of points on finite segment of line. Dedekind's cut are the different words about the same inconsistent notion: the infinite division of finite segment of line.

8. Geometry – usage rough physical models.

Morris Kline [2] is one of the mathematicians and historians of math who state problem of illogicality of mathematics. In 1985 [4, page VI] he written:

" ...for many vital phenomena, mathematics provides the only knowledge we have. In fact, some sciences are made up solely of a collection of mathematical theories adorned with a few physical facts."

Unfortunately, he did not notice that all geometry is the use of rough physical models for the expression of the abstract ideal objects that exist in the imagination of man only. It is very useful to show geometrical interpretations of math object, but initial notions of geometry have inside contradictions.

Initial notion of geometry.

Per Euclid [29, page 153], initial notion of geometry are:

"1. **A point is that which has no part.**
2. **A line is breadthless [without breadth] length.**
3. The extremities of a line are points.
4. **A straight line is a line which lies evenly with the points on itself.**
5. **A surface is that which has length and breadth only.**
6. The extremities of surface are lines.
7. A plane surface is a surface which lies evenly with the straight line itself."

From (3, 4) postulates we can see that point is part of line. If it is so then we can redefine point as segment of line which does not contain any point, because first postulate: "A point is that which has no part," and line is object which does not contain any points, because again point has no part. Inconsistency this definition is obvious.

Well known how much time was spend for fifth postulate of Euclid, but was not pay attention to inconsistency of notions: point, line and plane. From inconsistency of point and line follow inconsistency of notion of real numbers. To be clear, it is inconsistency to accept infinite division of finite segment of line: we cannot recognize infinite points of line as countable set, and all the more so uncountable of set of points.

Non-Euclidian geometries –Lobachevski, Riemannian geometry – are game with Fifth postulate, they are next level of abstractions, even more exactly, are redefinition abstract notion of strait line and parallelism of strait lines. For example, spherical geometry does not have strait line – there is geodesic line similar to strait line in sense as the shortest distance between two points – it is meridian. But the strait line and meridian has very huge difference – radius of curvature of the strait line equal infinity, and radius of curvature of meridian is finite. Meridian is not strait line. On the sphere we do not have parallel meridians, we do not have parallel shift of meridian – superposition of meridians can be done by rotation of meridian only. So we have here "abstraction-generalization" of the strait line and parallel shift to curve line and its rotation.

We can see sequence of steps:

Real hard body

Strait line and parallel shift

Curve line and rotation

The faith on reality curve space

Einstein's relativity theory

Experimental conformation of existence of curve space.

Geometry plays invaluable role in liveliness of abstract notion: notion which cannot exist because it is abstract, because it is ideal, because it has inside contradictions.

Geometry plays a particular role in mathematics and physics. Geometry is an abstract science with objects obtained from the abstraction and idealization of properties of solid bodies. Geometry has given birth to a lot of different kinds of abstract objects and made simple use of the following objects: irrational, rational, complex numbers and so on. The unification of the geometry of a solid body on space yields us more problems than advantages. The main problem is: space does not interact with any body or substance; this is the main property of space. Ivchenkov has shown [12] that Eddington's observations were within measurement error bounds. We still do not have any physical observations of any interaction with space. Geometry has within itself an inconsistency; namely, there is a point as an object without parts, as a geometrical zero.

Below three figure which illustrate overestimated "power" of projective geometry.

Figure 1. Geometrical "equivalence" of intellect of Mouse, Cat, Man and Tree.

Here "equivalence" of intellect is used as metaphor: we see the same angle of view on heights of mouse, cat, man and tree. It is not mean that they have the same physical height. All of them have different number of the same organic molecules, and the same distance between these molecules.

On Figure 2 we see similar triangles ABC and ADE. In math – not only in geometry, but in set theory, differential geometry, topology, and so on – is accepting that by drawing projection – line AFG, we can establish one-to-one correspondence between points of segment BC - F and points of segment DE - G, and if so then BC equal DE. In the same time we can make superposition BC on DE, and see that BC = B'C' = HE, or BC equal to part of DE - HE. This is result of accepting infinite division of the finite segment of straight line: part equal whole, and part equal part of whole. Logic could be retired.

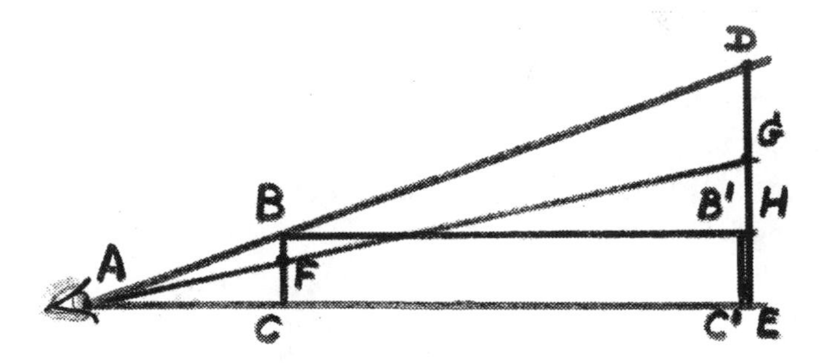

Figure 2. Geometrical equivalence of part BC and whole DE, and equivalence of part BC and part HE.

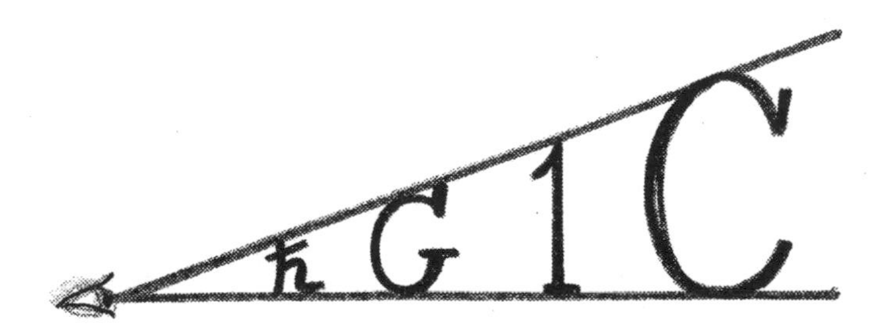

Figure 3. Geometrical "proof" of the absurd equation: $\hbar = G = c = 1$.

On Figure 3 we see physical constant: ħ [J*s] – Plank's constant (ħ = h/2pi), G [N(m/kg)^2] – gravitational constant, c [m/c] – velocity of light, and dimensionless unit 1 [32]. Usually to write this equation enough to say short mantra: "by choice of a system of reference." Here can not be omitted: those constant have different dimension, and different value. Convenience of this equation cannot be overestimated: these constants are used as multipliers, and equivalence to the dimensionless unit means that we remove them from equation, or we remove dependencies from these constant. Obvious it is case A: What do you wish: dependency or independency?

Projective geometry give us illusion possibility to create equivalence small and big segment of line, or finite area inside circle with infinite area on outside circle, or finite surface of sphere with infinite flat plain.

Inconsistencies Euclid's geometry begins with initial notion of "point without parts," line with length only, with plain without thickness and so on. This is corresponding with inconsistency real numbers – we cannot divide finite segment of line on infinite set of points, neither on countable set nor uncountable set. It is not enough even infinite expression on finite set of symbol to express uncountable set of real number. Notion of uncountable set was establish by using inconsistent notion of the infinite countable set in so called "diagonal argument." From physics point of view, we have finite set of natural and rational numbers.

All inconsistencies projective and Euclid's geometries go to non-Euclid's geometries, spherical geometry, and different types of Lobachevski's geometries. These geometries have its own inconsistencies: redefinitions of notion "straight line" and notion of "parallelism." A curve line is called as straight because it has similar characteristic – to be the shortest distance on surface, but flat and curve surface are very different type of surface. If flat has zero radius of curvature then the curve surface has some finite radius.

Notion of "parallelism" has meaning only for flat surface, because properties of parallel straight lines to keep the same distance between them, and we can not satisfy this for curve surface. For example, on sphere all "straight line," a meridian intersects each other and has different distance to next meridian. We can build line on sphere which has equal distance all its points to equator – it will parallel but not "straight line" like meridian.

Physics is under shadow two myths: myth about mathematical Nature, and myth about geometry which describe space. Geometry was born from abstraction of a solid body shape, than was artificial transfer on physical space, and was used as base to bend space like making deformation of solid body. From our point of view, any geometry does not exist in reality because it abstract and has inconsistencies. We can use geometry as approximation of a shape description. Existence of abstract theory of geometry can not be taken as basement for "discovering" properties of the physical space. It is main huge problem for mathematics and physics: "revive" abstract notions.

9. Paradoxes: Banach – Tarski: paradox, theorem.

Problems of development of mathematics and physics associated with different kind of paradoxes. It being understood that we should search a solution these paradoxes. But paradox is a consequence, and reason is an initial assumption or an undefined notion.

Well know math method "by contradiction": make start assumption, make several logical steps, and get result: if result is absurd, then start assumption was wrong or false.

Let's consider well known Banach-Tarski paradox, sometimes called as theorem [30]

"The Banach–Tarski paradox is a theorem in set-theoretic geometry, which states the following:

Given a solid ball in 3-dimensional space, there exists a decomposition of the ball into a finite number of non-overlapping pieces (i.e., disjoint subsets), which can then be put back together in a different way to yield two identical copies of the original ball.

Indeed, the reassembly process involves only moving the pieces around and rotating them, without changing their shape. However, the pieces themselves are not "solids" in the usual sense, but infinite scatterings of points."

Morris Kline writes [2, page 324]:

"A special case of this paradox [Banach-Tarski] discovered in 1914 is that a sphere's surface may be decomposed into two parts which can be reassembled to give two complete spherical surfaces, each of the same radius as the original sphere. These paradoxes, unlike the ones encountered in set theory of the early 1900s, **are not contradictions**. They are logical consequences of the axioms of set theory and the axiom of choice."

It is paradox but is not theorem we can see from result: one sphere was converted into two spheres. This absurd result directs on inconsistencies its base: axioms of set theory, where we see axiom of infinity, and axiom of choice for infinite sets. Indeed, set theory consists axiom of infinity which declares existence of infinity set – existence of not-existing because notion of infinity has inside inconsistence. Next contradiction is acceptance that finite sphere and its finite part can be divided on infinite "number" of points – derivative notion of infinity – "infinity in small," infinitesimal. Well know that set theory has a lot paradoxes, but enough one Banach-Tarski "paradox-theorem" to take consideration of inconsistency of set theory and, especial, contradictions of infinity as notion. The Banach-Tarski paradox related to topology which based on notion of continuity and infinitesimal values – both notion means accepting of infinite division of finite geometrical object. This paradox proves inconsistency initial notion of continuity and infinitesimal, and as well results of topology and differential geometry.

10. Mathematics and Physics.

Infinity generates derived notions which have inconsistencies too. There are notions infinitesimal, continuity, limits, and neighborhood and so on. All of them needs acceptance an infinite division of finite object.

Almost all branches of mathematics which is used in physics used these notions.

There are sources of inconsistencies which do not associated with infinity, but obvious have abstract nature and cannot exist in reality. They can be used as artificial method of manipulation by symbolic expression. It is mention above notion of zero – sometimes as empty space, or sometimes as number. Next inventions are negative numbers, which look like contrast with positive number, and usage together positive and negative number create illusion that contradiction can coexist. Complex numbers do not exist in reality; they are not number or are some kind of semi number, associated with number by list of rules. These rules of operations complex number with regular number give us power operate at least by three qualities: pure complex numbers, mixed complex-regular numbers and pure regular numbers. Illusion of calculation of quality automatically is overestimated in mathematics and physics. There is power method on the contrary: abstracting from quality, generalization of quality, or reduce or deny quality. It is all about group theory. By definition group, operation between two or more elements of group should give as result again element this group. In physics we can find operation with three different qualities, for example, calculation of impulse:

$$m[kg] \cdot v[m/c] = p[kg \cdot m/c];$$

Here quality of value associated with dimension of value, which show relation value to reality.

Mathematics and physics need review their tools – notions, concepts and theories on subject of inconsistency and illogicality. Before and now, on our point of view, they were not enough reflexive, prefer generate something different without analysis on inconsistence and logic. Well known: "Habit dies hard."

11. Logic and Logics.

Logic is very important and complex subject. Here we will emphasize several aspects of logic which is related, from our point of view, to discussed topics of the book.

First of all, we need logic for deductive reasoning, to get correct conclusion from correct data or statements, to estimate correctness of initial assumptions and information. Logic is a necessary part of scientific method which should prevent to accept illogical statements and inferences.

Well known fact is that deductive power of mathematics and any science is based on logic. But, on other side, man loves paradoxes, and takes his intuition, especially in math, above logics. It suffices to recall Bohr's words: **"Your theory is crazy, but it's not crazy enough to be true.,"** and his complementary principle – "contradictions do not contradict but complement each other."

It is impossible overestimate a role of logics in science. Logic is base of science in contrast to base of art and religion – authority and a faith. Main quality of science is reproducibility of prediction on the natural phenomena.

Main goal of art is creation unique artwork, but not good copy anything. And it will be bad photo camera which will give as each photo with unique unpredictable effect. Art needs good authority and the faith to him.

Religion or religious activity is based on Authority and the faith to Authority and religious dogma.

Science, from our point of view, should be distinguished from art and religions by usage logic and reproducibility of logical conclusions.

Nonetheless, there is a theoretical physicist who attempts combine metaphysics, physics, theology to theosophy as the highest level of knowledge, for example, by Vladimirov Y. S. [31]

Logic is one of tool – with initial notion, axioms or postulates, to compress knowledge, and create "new" knowledge by "logical conclusion." But we have lot logics. There are logics which we should not accept, because they destroy the actual logic and they make visibility of logicality of the illogical concepts and conclusions.

It is dialectic which accepts mutually eliminating notion and concepts, and even more, dialectic declare "coexistence" of these concepts as law of Nature. Karl Popper criticizes dialectic in his nice work "What is dialectic?"[33].

> Even in the mathematical logicism by Frege we can see completely illogical statement which declares existence of not existed. Here is declaration by Frege notion on zero [34, §74]:
>
> "Since nothing falls under the concept "not identical with itself," I define nought as follows: 0 is the Number which belongs to the concept "not identical with itself." ... All that can be demanded of a concept from the point of view of logic and with

> an eye to rigor of proof is only that the limits to its application should be sharp, that it should be determined, with regard to every object whether it falls under that concept or not. But this demand is completely satisfied by concepts which, like "not identical with itself," contain a contradiction; for of every object we know that it does not fall under any such concept."

We see, that for the sake of logic and rigor of proof Frege involves concepts having contradictions, and it defines an object, which does not exist, because "we know that it does not fall under any such concept." We can find a lot contradiction and illogicality in mathematics and physics, but we should not accept them, we should draw the right conclusion. Discrepancy of notions of infinity, infinitesimal, and zero leads to the inconsistent notions of irrational numbers, continuity, and geometrical objects. Applications of these notions do not prove their consistency, and they can be estimated as approximations.

12. Math method: abstraction, idealization, generalization.

Abstraction and idealization create problem: abstract and ideal object "leave" in man imagination only, and does not exists because its own inconsistencies.

Point is physical object which size we can ignore in context of our task. Line with one dimension, plane with two dimension, sphere and so on do not exist because they are abstract and ideal.

Core of math methodology: abstraction and idealization – give us huge problem like Uroboros – snake which bit itself for own tale.

Usage undefined concepts and axiomatic method in mathematics give us a lot mathematics and Morris Kline proposes make a choice from multiple math on our own test. The faith in some set of axioms is religious activity not science activity.

In geometry a long time used accustomed notions of point, line and plane, but those notions have inside contradictions and because this cannot exist. Points, Lines and planes exist in mind and imagination of man only.

Well known that some abstract notion appears in mind of different people, but from this is not follow conclusion about existence this notion independently from man. People have the same ability to make abstraction, and they observe the same Nature.

First abstraction itself is leaving from real object, not way to reality like per Penrose [11], and next levels of abstraction can make "distortions" and "deformation" of content and form of real object. It is reason for conclusion that intuition and imagination independent from reality. An ideal abstract mathematical object can have esthetical and emotional assessment: it is based on limited abilities our senses and our devices, because we evaluate real object which is close to mathematical object – symmetric crystal or snowflake. We use very rough physical models for point, line and plane, but admire the Pythagorean Theorem.

Men senses and devises are limited in his/its abilities, but it is not mean that invisible is not real, that invisible objects or phenomena are not exists. We can operate invisible object by usage visible one: using visible pump we can put invisible molecule of air to soccer ball. Different case is what conclusion we can do: we can move air to ball, we can feel warming of pump, but we can not decide that air is mix of nitrogen and oxygen. A much more complicated problem is to operate with invisible part of invisible object, and, first of all, we should distinguish invisible object and its part from our abstract notions.

Usage mathematical methods give us abstraction of invisible object, and "existence" this abstraction "confirmed" by visible experiment and visible result. One of physical axiom could be statement that abstract object does not exists, exists physical object and related to it abstract notion. We have notion of "atom" and this notion confirmed by millions physical and chemical objects and phenomena. Abstract notion of atom does not exist in reality: exists atom of hydrogen, atom of oxygen, atom of carbon and so on. Hegel (1770-1831) had mentioned that we can eat cherry and plumbs, but cannot eat fruit. Even more: we do not eat word "cherry," we eat physical cherry. One opponent had mention that any word can be printed on rice paper by eatable paints and can be eaten. But it is not cherry and dangerous for health of opponent.

Example: electron is wedding of two abstract notions: atom and charge. We should get back to Thomson and Kaufmann experiments.

Well, we cannot hope that mathematicians will be more reflexive, especially, pure mathematicians. They prefer to drink with Godfrey H. Hardy (1877 – 1947), for his toast: "Here's to pure mathematics! May it never have any use."

And well known that there is Freedom of Thoughts, pure mathematicians have it too, but with a little note: drink for any toast, but nether for public expenses, nor for taxpayer's money.

All abstract notions, concepts, axioms, theorems, theories do not exists in reality. All of those exists in man's mind and in form of paper's expression as approximate model.

Pythagorean Theorem is approximation.

Do not exit pi, e, square root from 2, square from minus 1, real, transcendental, irrational numbers, exist natural number, and rational as expression with natural numbers.

Mathematical Nature is myth came from past history. In reality mathematics is tool to create approximate model or description of reality.

13. Physical "axiom"– reality of physical object.

Roger Penrose, in his "The Road To Reality. A Complete Guide to The Laws of The Universe" write [11, page 1027, 1028]:

"Yet, some readers may well still take the view that the road itself may be a mirage. True – so they might argue – we have been fortunate enough to stumble upon mathematical schemes that accord with Nature in remarkable ways, but the unity of Nature as a whole with some mathematical scheme can be no more than a 'pipe dream'. Others might take view that the very notion of a 'physical reality' with a truly objective nature, independent of how we might choose to look at it, is itself a pipe dream."

Usage of inconsistent mathematics definitely gives us a 'pipe dream'. Any notion is man made concept in man imagination. The notion of physical reality corresponds with real physical object which exists independent from our notion about that. This physical 'axiom' does not need to be proved by throwing a stone to head of opponent. We should avoid usage the second law of research institutions in this case: it is simple task – reality exists.

Steven Weinberg, in Chapter VII Against Philosophy, writes [35, page 167]:

"Physicists do of course carry around with them a working philosophy. For most of us, it is a rough-and-ready realism, a belief in the objective reality of the ingredients of our scientific theories. But this has been learned through the experience of scientific research and rarely from the teachings of philosophers."

From physics point of view, here we can see two problems:

First of all, "ingredients of our scientific theories" are not the objective reality.

And secondly, any activity, especially scientific research, can not learn about existence of abstract notion. We cannot eat fruit, but can eat apple, we can not prove existence of fruit, independent how much apple we will eat.

Part 2. The Search For Physics.

From historic point of view, now we can consider physics as combination of mathematics, natural philosophy and experiments or observations. Mathematics and philosophy are not enough reflexive, have problem with logics, very easy generate notions and concepts without analysis of inconsistencies these notions, and manipulate notions to "prove" logicality of the illogical notions by experiments and observations.

Physics is not mathematics and it is not philosophy. Physics is science which study Nature to get knowledge about Nature. As science physics should be useful activity – point of view Leo Tolstoy, but not activity which should proof power of man's intellect. Intellect one man or men's community can make mistake, and keep this mistake as intellectual achievement.

Steven Weinberg describes [35] very enlightening case of discovering of electron by J.J. Thomson. He compares work of Thomson and Kaufmann. He writes with irony or even with sarcasm [35, page 179]:

"In retrospect the positivism of Kaufmann and the opponents of atomism seems not only obstructive but also naïve. What after all does it mean to observe anything? In a narrow sense, Kaufmann did not even observe the deflection of cathode rays in a given magnetic field; he measured the position of a luminous spot on the downstream side of the vacuum tube when wires were wound a certain number of times around a piece of iron near the tube and connected to a certain electric battery and used accepted theory to interpret this in terms of ray trajectories and magnetic fields. Very strictly speaking, he did not even do that: he experienced certain visual and tactile sensations that he interpreted in terms of luminous spots and wires and batteries. It has become a commonplace among historians of science that observation can never be freed of theory."

1. Last statement of this citation – "that observation can never be freed of theory" - is completely wrong exactly from historic point of view. To prove that we can consider time-line of technology and time-line of development of theory. For example, mankind use fire about one and a half million years, steam engine was used about 2000 years, James Watt developed steam engine in 1763, Russian inventor Polsunov had built his steam engine in 1764, but "father of thermodynamics" Sadi Carnot had written his work "Reflections on the Motive Power of Fire" which started thermodynamics in 1824. Irish-born British physicist Lord Kelvin was the first to formulate a concise definition of thermodynamics in 1854 [36]:

"Thermo-dynamics is the subject of the relation of heat to forces acting between contiguous parts of bodies, and the relation of heat to electrical agency."

2. Problem was not and is not in positivism or atomism, like is not in any "–ism" of philosophy. Atom is physical notion about substance, but it is not exists in Nature – exist Hydrogen, Helium, Lithium, Beryllium and so on. Physics begins from "luminous spots and wires and batteries," and notions and theories appear after beginning. The deep problem of physics is misuse of "prediction" some theory as "discovering" of existence some abstract notion by experiment. History of physics has many examples of this problem: ether, field, electron and so on. Below we will consider this problem in detail.

First level has a small list of notions: Universe, Nature, and World. We do not need to resort to undefined notions to begin and then proceed; we already have notions which correspond to reality.

The next level includes substance, space, interaction, no interaction, dependency, and independency. The notion of matter we propose leave for the philosophical body-mind problem. We accept the independency of reality from our mind. The 3-dimesional substantial body moves in the 3-dimentional space. Substance does not interact with space. Two substantial bodies can or cannot interact with each other. That there are no interacting bodies or independent bodies or bodies which are interacting which can be ignored, this is the core of Galileo's principle and Newton's first law. Bodies can be independent because of the inverse square of distance law: interaction can disappear and emerge again – this is base of probalistic behavior of moving bodies. Substantial bodies may have internal structure or sublevels having the notions: substance, space, interactions (atom or molecular level).

The next level has the notions time, distance, velocity, energy, mass, and so on. It is important to note that time is the man-made notion to describe motion of body in space, and cannot be mixed with space and substance.

The next level may include emergent, system notions like temperature, entropy and so on, which bind with the internal structure of a body and do not make sense, for example, for one molecule.

The shining example of the mixing of notions of different levels of abstraction is proposed by Hawking to discover new spatial dimension by holography [37]. It looks like 2-dimensional photo plate generates 3-dimensional image, but in reality 3-dimensional micro object - photo plate, generates 3-dimensioanl macro image.

14. What is Physics?

It may seem to force an open door to search answer on this question. It is very simple to find a lot definition of "physics" and "science of physics." There are several examples of definitions of physics. In The American Heritage Dictionary of the English Language [38] we see:

1. "The science of matter and energy and of interactions between the two, grouped in traditional fields such as acoustics, optics, mechanics, thermodynamics, and electromagnetism, as well as in modern extensions including atomic and nuclear physics, cryogenics, solid-state physics, particle physics, and plasma physics."
2. "The study of the natural or material world and phenomena; natural philosophy."

In "PHYSICS. PRINCIPLES WITH APPLICATION," by Douglas C. Giancoli [39, Page 1]: "Physics is the most basic of the sciences. It deals with the behavior and structure of matter. The field of physics is usually divided into the areas of motion, fluids, heat, sound, light, electricity and magnetism, and the modern topics of relativity, atomic structure, condensed-matter physics, nuclear physics, elementary particles, and astrophysics."

[39, Page 2]: "The principal aim of all sciences, including physics, is general considered to be the search for order in our observations of the world around us. Many people think that science is mechanical process of collecting facts and devising theories. This is not the case. Science is a creative activity that in many respects resembles other creative activities of the human mind."

[39, Page 4]: "For a long time science was more or less a united whole known as natural philosophy."

On Wikipedia [40] we read: "**Physics** is the natural science that involves the study of matter and its motion through space and time, along with related concepts such as energy and force. More broadly, it is the general analysis of nature, conducted in order to understand how the universe behaves."

From Aristotle – Galileo – Newton physics can be considered as the mathematical natural philosophy with experiments and observations. In modern physics special role plays mathematics: mathematics is language of science and physics; mathematics itself studies Nature, because Nature follows mathematical laws. Fact is that mathematics and physics is not enough reflexive as science should be. This issue was lost in history of math and physics. Morris Kline bring this problem for mathematics, but find solution this problem in fruitful and useful application of math in physics, and physics confirms this fruitfulness of math by experiments – here vicious circle was closed and self-reflection of science died.

First of all, in math and physics should be removed contradistinction of mind and senses. Let's try to take a look on history and call a spade a spade: all of them – senses and mind as any kind of intuition – can make mistakes, and we should tested for different condition our knowledge, with combination senses and mind, to get reproducibility.

Secondly, math as tool for physics should be reviewed on illogicality and inconsistencies of notions, concepts, axioms and theories.

Third, physical methods or theories should be tested by physical experiments to avoid declaration of great "discovery" by one kind of interpretation of some experiment. Never late to walk back to repeat experiments, and review and rebuild theory. Several "steps-back" will be consider below.

15. Substance – space – motion.

Physics is science where is studied Nature around us. We have physical objects and phenomena, and symbolic knowledge of those realities. Symbolic knowledge is result of study Nature, but as result of study can be consider device or technology too. Nature of any notion is abstraction from Nature, physical objects or phenomena. Abstraction is product by man and for man. One of kind of abstraction is symbolic form.

Symbolic knowledge should have initial notions or concepts. Initial notions can be organize in hierarchical system to reflect some kind "distance" between notion and reality, and dependency one notion from other one. Let's propose here next initial notion:

… (Nature itself) …

Multiple physical objects and phenomena

Nature

Substance – space – motion

We can say: Nature is substance which moves in space. It seems that notion of matter could be kept for philosophy to express independence of substance-space-motion from man's mind. From this point of view – Nature, substance, space, and motion are completely material.

Not acceptable expressions are:

Space moves in space,

Space moves in substance,

Substance moves in substance on atom or molecular level. On macro level we observe diffusion, dissolution, absorbing and so on. Here was mention notion of atom and molecule as the smallest part of substance which cannot occupy the same space, or atom cannot moves in other atom.

Next notions express relations between mention above. There are:

Interaction, not – interaction, dependence, independence.

Substance can interact with other substance: it is case when one substance can change motion of other substance. This interaction expresses dependence one substance from other, or dependence motion one substance from another substance or its motion.

We can observe not-interaction between two or more substances. It can happened because distance between body, or interaction between substances so small that we can ignore them.

Substance and space are not interacting with each other completely. Existing concept interaction of space and substance based on experiments with bad metrology [12], they results lay in range of error of measuring, and they was defended by authorities – beginning from Eddington to Einstein and followers of relative theory.

Very important note that initial concept is absolute independence of space from substance: all observable interactions are interactions between substances. Experiments of light's deflection should be analyzed from point of view metrology. More arguments independency space and substance will discuss below.

Important notion of independency will consider in section Time. Here we can note that in unlived world we observe independent objects and phenomena: from absolute independency to dependency which we can ignore. In biological lived world we see fundamental concept of cell – part of organism which independent from other cells and around environment. Cell is visible expression of concept of independency. There are often used words "as whole," and understanding this as principle "all interact with all." It is wrong point of view, because from system point of view part of system can depends from one part and be independent from another. Max's principle is overestimation of dependency and interactions, and it is underestimation of independency and existence of not-interacted objects.

There is mathematical myth that geometry studies properties of space. Geometry is studied abstraction which was made from substantial hard body, and ideal geometrical abstraction "exists" in man's mind only, but nowhere more because they have inside contradictions like point without of parts, line, plain, sphere and so on. To work with abstraction in geometry we need to use rough physical model which content a lot molecules even in "point" without parts.

16. Dimensions.

Dimensions are next level of abstraction. All physical object are 3-dimensional, no dimensionless, one or two dimensional objects. We can consider dimensions as three especial bodies which we used to describe positions and motions other physical body. Mathematical dimensions we get as abstraction from these auxiliary bodies. We ignore mass and two dimensions these body for convenience of description of motions. In this case we associate dimensions with geometrical line which as abstract ideal object does not exists in reality. In general case dimension is some kind of characteristic physical body, phenomenon, or description, or argument of function.

Next notion tied with dimensions is direction of orientation or direction of motion. Dimension as straight line links to notion of orthogonality – one kind of independency of dimensions and directions. We can see or show three space dimensions and three orthogonal directions. In this case we can use Pythagoras's theorem to calculate distance. Attempting to add more dimensions broken orthogonality and usage of Pythagoras's theorem.

Problem of the multiple space dimensions – more than three, is problem relation of abstract world with real world. Additional dimensions in existed theories are invisible in reality, can be tested indirectly and depend from interpretations.

Physical approach to notion of time we will consider in next section.

17. Time.

There are several approaches to the problem of Time [41-48]:

1) Unreality of time and motion. Parmenides (510-450BC), Archimedes (270-212BC), Lucretius (70BC), McTaggart (1908,1927).
2) A-theory (past, present, future), Presentizm, exists only NOW [43].
3) B-theory (earlier, simultaneous, later) [43].
4) Newtonian absolute time, separated from space [41].
5) Time as one of the dimensions of space-time: 3+1, 4+1, 9+1, 25+1, …
6) Hypothesis of two kinds of time (Itzhak Bars) [45].
7) Hypothesis of multidimensional time (Pavlov) [46].
8) Composite hypotheses. A-B-theory [43], and the hypothesis by Julian Barbour, 1999, includes three approaches – (1) "time does not exist at all"([44] p.4),(2) "The world is made of Nows"([44] p.16), (3) "unification of general relativity and quantum mechanics may well spell the *end of time*." ([44] p.14).
9) Reality of time by Lee Smolin [47] 319p. "the most scientific is reality of time."
10) Substantial time by Levich [48].

In this work we attempt to propose a different approach to the problem of time on the base relations of dependences and independences of moving into space body.

Let's consider an example. Suppose we have to describe motion of a car. To do this we need to measure distance and time which car spent to move on this distance. To measure the distance we can use a wheel of the car and assume that circumference of wheel is equal one meter. To measure time as a rule use a periodic process and this process is convenient to get numeric value of duration of time. If we look at the car then we can make a note that the rotation of wheel of the car is a periodic process and in this example let's use it to define one turn of the wheel as "one second." We remember that length of circumference is equal one meter and we can measure turns of wheel in meter too. Thus, we can express the velocity of the car as a ratio of "one meter" of distant to "one meter" of time (turn), and in this case the velocity of the car will equal dimensionless unit independent of the real speed of the car. Here we have time with at least two benefits: time is very local, and time has the same dimension like space.

This example is necessary to formulate the reason why this way to measure of time cannot be accepted. The reason is a complete dependence of rotation wheel upon the motion of the car, and that does not matter: car moves because the wheels rotate or the wheels rotate because car moves. When the car is still – motionless wheels, when the car accelerates – accelerate the rotation of the wheels. Therefore, to measure time, we need a periodic process independent of described process. We could consider a pendulum, mechanic watch, electronic watch, atomic clock, or using outside periodic radio-signal of artificial or natural sources. The more independent clock is the more accurate to measure of the speed of the car. Hence it follows definition one important properties of time as relations of independency at least two the moving in space processes, one of them we call clock. If we try to establish dependences to time, for example, stop growing old or travel in time, we get nothing more but broken clock; in other words, we can always find an independent process or the clock which we still did not break. Any process can not be used as a clock to describe itself, and any clock cannot be used to describe its own motion.

We have, observed and experimented with three-dimensional matter moving into three-dimensional space, where exists a relation of dependences or cause-effect relation, and a relation of independences which is bound with the notion of time. Subjective role of time or time as man-conception is necessary for us to describe the motion of object, where the clock is independent of the described motion and the described motion is independent of the clock, and motion owes its cause, and the cause is independent of clock and our description. The concept of time is necessary as a standard motion to get a description of other motion, so time is derived from motion – substance, space and motion are primary notions, and time is derived concept.

From example above, the description of motion of car with wheel-clock shown the same notion of velocity and acceleration of car as four-velocity of special relativity theory: per [13] four-velocity equal unit and orthogonal to four-acceleration. Both of them are physical absurd.

In respect to Newtonian time [41] we have to note that using time separate from space completely satisfies reality and concept of time as standard motion.

If we will follow physical approach then we should see, first of all, physical object. We considered in Section 16. Dimensions three body as base of abstract notion – dimension. The abstract notion of time we can associate with body which has standard unified periodic motion, and this motion is convenient correspond to number; number of time is expressing duration of standard motion.

Important to notice that when we took three bodies for dimensions and one body for time, we do not change any properties of substance-space-motion. We make abstraction of dimensions and time without request to change dimensions of space and dependency of time.

Let's take a look on different approaches to notion of time: timeless, illusion of time, real time, substantial time. All of these approaches follow steps to main problem of physics: revive abstract notion accordingly to the Weinberg philosophy with the faith of existing "the ingredients of our scientific theories."

Per physical approach we can see that:

Timeless - YES, because Nature does not need notion of time, NO – because all physical process has duration, and "duration" cannot be separate from process, we can separate the notion of time only.

Illusion of time - YES, because of abstract notion of time does not exist, NO – because the motion of substance in space is real.

Real time – NO, abstract notion does not exists, YES – exists the physical process with duration of motion.

Substantial time - NO, abstract notion cannot be reified by any axiom or theory, YES – because body of standard motion which we call a clock, is substantial.

Thus, main problem of physics looks like steps:

Physical object or process

Abstract notion associated to physical object

Abstract notion from other notion

Reifying or revive abstract notion

Experimental discovering revived notion.

All these steps physics takes with the huge help of mathematics, but last step – experimental reifying of notion, is pure responsibility of physics.

18. Mass.

In physics notion of mass has several meaning. We can find in the Maxwell's treatise [49,50] the definition of mass.

From physical point of view: " …all masses ought, if possible, to be compared directly with the standard, and not deduced from experiment on water." [49, p. 3, 4].

From mathematical point of view: " ...masses themselves to have no other mathematical meaning than the volume-integrals of $\frac{1}{4\pi}\nabla^2\Psi$, where Ψ is the potential." [50, p. 177].

Russian physicist V. Fock distinguished two kinds of masses: inertial and gravitational. He had noted that those are different mass, but proportional only [51, page 152]. The Equivalence Principle declares that inertial and gravitational masses are equivalent. It is considered that from time of Galileo "all matter responds to the gravitational field in exactly the same way regardless of mass or internal composition."[52] "Regardless of mass" – it is half of truth. By Newtonian law of gravitation:

$$F = G\frac{M \cdot m}{(r+h)^2};$$

Where h is height of position of body with mass m. M is mass of the Earth, and r is its radius, 6370 km. Thus acceleration of mass is:

$$\frac{F}{m} = G\frac{M}{(r+h)^2};$$

Here we see that different mass falls with the same acceleration because get action of different force F which is proportional to this mass, and acceleration is variable, it depends on the height h. This dependence is hard to measure because height is too small related to radius of the Earth, 6370 km.

Most of scientists take one kind of mass which participate in inertial and gravitational phenomena. Nevertheless, this one kind of mass has four different treatments:

A) Newton's mass as absolute value in the sense independent from inertial system of reference;
B) Relativistic mass of Einstein's relative theory which dependent from velocity of body;

C) Absolute mass in the Einstein' relative theory by Lev Okun [53], which independent from value of velocity of body, as Newtonian mass.

D) Relativistic mass by Lev Okun [53], which dependent from impulse or direction of velocity.

Most of physicists accept case B), but Lev Okun is expert in the relative theory, and belong to small group of physicists into not big community of contemporary physicists who study particle's physics, and only physicists this small group, per Okun, understand the Einstein's theory of relativity.

Nobel laureate, physicist V. Ginzburg has three points of view on treatment of mass [54]:

1) He agrees with L. Okun "in the methodological, pedagogical and historical relations";
2) He notes, that the choice between dependence or independence of mass on velocity seems a subject of "taste" choosing "and argue there is nothing to";
3) It is not clear for him "Where and when to stop in such disputes".

Okun's position is based on followed formula to define mass [53]:

$$E^2 + \boldsymbol{p}^2c^2 = m^2c^4$$

For photons $m = 0$, and formula $E_0 = mc^2$ we get when $\boldsymbol{p} = 0$. But impulse in classical and relativistic cases dependents from velocity $\boldsymbol{v}$:

$\boldsymbol{p} = m\boldsymbol{v}$ Classical impulse;

$\boldsymbol{p} = \frac{\boldsymbol{v}E}{c^2}$ Relativistic impulse.

It is contradicting to the Okun's point of view.

Max Jammer, after consideration of Okun's works, writes [55, page 61]:

"Our analysis of the m [Newtonian mass] vs. m_r [Relativistic mass] debate thus leads us to the conclusion that the conflict between these two formalisms is ultimately the disparity between two competing views of the development of physical science."

Let's notice that there is disparity of the different treatment of notion of mass after more then hundred years of development the theory of relativity. In Section 23 we will consider initial inconsistencies of SRT and GRT.

19. Maxwell's equations.

First of all, we should emphasize that Maxwell had consider in his Treatise [49, 50] electromagnetic medium, but not ether. We can find often reference as "Maxwell's ether." He writes [49, page 68]:

"That the energy in any part of the **medium** is stored up in the form of a state of constraint called **electric polarization** ..."

"the dielectric medium, ... even what is called a vacuum"[49, p.68, art.62]

And further in [50, page 431]:

"We have now to shew that the properties of the **electromagnetic medium** are identical with those of the luminiferous medium."

Maxwell involves notion of electric displacement and notion of electric currents [49, page 65]:

"The variations of electric displacement evidently constitute electric currents."

Maxwell [49, page 76] write electric displacement as:

$$D = \frac{1}{4\pi} kE \qquad (19.\ 1)$$

Here $E = \frac{q}{r^2}$. Now it is very important to notice that electric polarization and electric displacement keeps dielectric or medium locally electrostatic neutral. It means that D is not function of one kind of charge, but E is function of one kind of charge q. Left and right sides of equation above are inconsistent.

The same notice we can do for electric current in conductor: it is electrostatic neutral until molecular level. Notions of a volume charge, density of charge, current as flow one kind of charges is inconsistence without consideration interaction of one kind of charges. Or there must be not charge, but some object without repulsion of each other. Maxwell's equations consist two functions E(q) and H(j), where j is density of current which inconsistent with density of charges q. Thus, the modern Maxwell's equations are inconsistent.

On any a physics notebook we can learn the wonderful electromagnetic unity of Nature – any substance consists two kind of charges, negative electrons and positive kernel, which participate in electromagnetic interactions, and we see all these by the electromagnetic waves – light. The base of this unity, specially the electromagnetic nature of light, is the Maxwell's equations [13, 49, and 50].

Let's consider the differential modern form of the Maxwell equations [13].

$$curl\,\boldsymbol{E} = -\frac{1}{c}\frac{\partial \boldsymbol{H}}{\partial t}, \qquad (19.2)$$

$$div\,\boldsymbol{H} = 0, \qquad (19.3)$$

$$curl\,\boldsymbol{H} = \frac{1}{c}\frac{\partial \boldsymbol{E}}{\partial t} + \frac{4\pi}{c}\boldsymbol{j}, \qquad (19.4)$$

$$div\,\boldsymbol{E} = q, \qquad (19.5)$$

There is well-known mathematical proof of existing of the electromagnetic wave in source-free space [13, p.116]:

"The electromagnetic field in vacuum is determined by the Maxwell equations in which we must set $q = 0, \boldsymbol{j} = 0$. We write them once more:

$$curl\,\boldsymbol{E} = -\frac{1}{c}\frac{\partial \boldsymbol{H}}{\partial t}, \qquad (19.6)$$

$$div\,\boldsymbol{H} = 0, \qquad (19.7)$$

$$curl\,\boldsymbol{H} = \frac{1}{c}\frac{\partial \boldsymbol{E}}{\partial t}, \qquad (19.8)$$

$$div\,\boldsymbol{E} = 0, \qquad (19.9)$$

These equations possess nonzero solutions. This means that an electromagnetic field can exist even in the absence of any charges.

Electromagnetic fields occurring in vacuum in the absence of charges are called *electromagnetic waves*."

We see here the simplest approach: deleting the explicit written in (19.2-19.5) charge q and current **j** and mathematical manipulation with the rest of symbols of the Maxwell equations.

But ***E****(q)* when $q = 0$ equals zero too: ***E****(0)* = 0, and ***H****(j)*, when ***j*** = 0, ***H****(0)* = 0. Thus it is looking like "proof" by mathematical tricks existence of zero in reality in form electromagnetic waves. Here we see "reviving" of zero with different abbreviations: ***E*** and ***H***.

20. Light.

Despite it may seem incredible, but question: What is light? - Still is open. We live into a light environment, and with light and through light: light gives us life and life on the Earth. It may seem that notion of light directly correspond to physical object or phenomena, but attempting to understand light, we had involved intermediate concepts. There are a lot of them:

Newtonian corpuscles,

ether where light is stress and waves into ether,

Maxwell's electrical medium where light is displacement and wave of charges of medium,

massless electromagnetic field,

massless Plank-Einstein's photons,

as energy which can arise and can disappear, or gives birth to particle of substance,

as wave-particle duality with relative or absolute speed,

as some kind of symmetry in manifold of different conceptions,

as morphism in defined math structure.

There still were not discovered a mass corpuscles, and medium like mechanical ether or electrical Maxwell's medium. Above was shown that mathematical trick with symbols of Maxwell's equations cannot be taken as proof of existing of a massless field or photons. As expert of SRT and GRT, Lev Okun [53] had discussed tenth years about massless energy of photon, and absoluteness mass in SRT in meaning independence of mass from velocity. The massless photon with the absolute velocity has the mathematical "basement" of existence of zero, existence of empty-not-empty place which can have any properties. In this case physics is pure mathematics. From our point of view, problem of light is initial notion and task to get answer on questions:

What is light?

What is propagation of light or phenomenon on distance?

What is velocity of light in case of propagation or a phenomenon on distance?

We cannot make a world view conclusion based on unknown light without answers on these questions.

21. Electrical Charges – surface phenomena.

Word "charge" has lot meanings, even in physics it has many meanings too: electric charge, color charge, magnetic charge. Quantum numbers can be considered as charges, which sometimes called as the Noether charge. Mass can be considered as charge of gravitational field. We will speak about electric charge as quantity of static electricity.

It is necessary to allocate followed problems of notion "electric charge" or "electrostatic charge":

1) Notion "electric charge" is physical notion and very wide used a long time as surface phenomena. But we still can not measure electrostatic charge: we have electroscope to estimate charge on quality level; we have digital devices

which measure current of discharges or current of induction of electrostatic charge. We still cannot mesure integral charge of body, and density of charges on surface of body.

2) In well known Coulomb's law we completely ignore phenomenon of electrostatic induction. We use charges in this law which was charged on big distance from each other. When we make them closer to each other electric charge of both of them will be changed by electrostatic induction. Thc Coulomb's law is very symmetric to the Newton's law of gravitation, but for mass we do not observe the induction phenomenon.
3) In electrostatic phenomena we think about separate kind of charges. In electrochemical phenomena we cannot separate those charges and can use both of them.
4) Usually an electric current is defined as a flow of electric charge. But an electric current is the electrostatic neutral until molecular level. Maxwell called electric current as kinetic electricity. In electrical phenomena as discharge in air or in liquid we observe electric current. The observed lightings have different curvature, and looks like Christmas garland: bulbs are not move, only illuminate in course to make illusion of motion of light. It seems in lightings is changed state of air in different place, and we see illumination of motionless air.
5) Long time we observed electrical phenomena with connection to notion of atom. Particularly, it was related to Faraday's electro chemical research. It was so close: notion atom and charge, that Thomson combined these notions in atomic charge and called it as "electron." In next section we will consider this notion of electron, because it still is question: is electron physical object or abstract notion as combination of abstract notions of atom and charge without correspondence to reality.

22. Electron.

Well, it had happened: notions of atom and charge were friends so long that we have now electron. But with electron we have many problems and questions. It looks like electron is not work even for the quality explanation of electrical phenomena.

Let's consider electrostatic phenomena. We have two neutral body and process of electrification by Maxwell. What should do electron in process of electrification of body? It must leave one neutral atom and move to neutral atom of other body. When electron leaves atom, atom becomes positive charged ion. Question is: what kind of force does move electron from positive ion to neutral atom? Next step of electrification of body put next question about force which moves electron from positive ion to negative ion, because neutral atom becomes negative ion after first step of electrification. Similar consideration can be taken for chemical reactions.

Ok, in 1913, Niels Bohr had killed if not electron itself then its relation to electromagnetic theory: electron should but don't illuminate on atom orbit. It illuminates when jump form one orbit to other.

Chemistry and quantum chemistry have its own electron with different behavior, electron moves in useful for chemistry direction, or looks like cloud of density of square of probability.

In experiments of diffraction electron can try to be in different whole of screen in the same time to help physicist explain picture of diffraction.

Thus we have one notion of electron and a lot set of different its properties, so different that irresistibly arises question: what is electron: real object or abstract notion with ad hoc usage?

From our point of view, this question is initial and very important, because it starts a long sequence of notions – from positive kernel of atom to Higgs boson which looks like abstract notion of particles' mass - mass was separated form particle and successfully discovered as real particle. Here we can note inference of mathematical method of abstraction, and physical power of discovering of reality with respect to the Weinberg philosophy.

23. SRT, GRT

SRT – it is the special relativity theory, GRT – it is the general relativity theory, or we will use simple the relativity theory.

Despite that the relativity theory is called now as classical physics, and old classical physics is called as before-relativistic physics, I guess, it well known that the relativity theory is illogical. There are international communities [56, 57] where can find a lot work with criticism of the relativity theory. Here we will mention about important, form our point of view, problems with the relativity theory.

SRT was invented as solution of the difficulties of Maxwell's electromagnetic theory. As was shown above, existence of electromagnetic field was done as mathematical trick. Using the problematical theory to change fundamental, world outlook concepts, it was, mildly speaking, boldly. The Michelson's experiments shown negative results related to expectations. On time those experiments it was unknown what is light, ether was artificial invention, and it was man's assumption of the way of interaction light and ether, interaction light and mirrors, interaction ether and substance. All of those man's dreams were not enough to make revolutionary postulate that velocity of light is absolute: independent from velocity of inertial system of reference. Before velocity was relative value. Relative and absolute values have completely different quality: they are incompatible and sum those value does not make sense. Relative velocity is measuring as distance per time. To use this formula for "absolute" velocity, it was involved Lorentz' transformation space and time: it is really - if Nature does not satisfy our theory, then it was worst for Nature; the mountain should go to smart man.

Next invention was combination of space and time as whole concept. As was shown above, time is man's concept, involved for the men's needs to describe motion of substance. The relativistic 4-dimenssional space-time gives us meaningless notions [13]:

4-velocity which is dimensionless unit,

4-acceleration which is orthogonal to 4-velocity, and that means that 4-velocity can change only direction, but not value, if 4-acceleration is not equal zero;

4-velocity of light has infinite components;

interval for light equal zero in all inertial system of reference, and interval for body which rest in inertial system of reference grows with velocity of light c.

GRT was based on the inconsistent non-Euclid's geometry and on interpretation of experiments by man's wishes. The inconsistency of geometry we discussed above. Experiments with deflection of light near the Sun was interpreted by Eddington and was accepted because Eddington's authority.

Ivchenkov has shown [12] that Eddington's observations were within measurement error bounds. We still do not have any physical observations of any interaction with space. We have interpretation of cosmic picture about a gravitational lens, and we do not have analysis of metrology even for observation where we can and should analyze metrology of measurement.

From our point of view, SRT and GRT are nice examples of reviving of abstract mathematical notions. We can find [58] that Einstein's merit was in accepting Lorentz's transformation as physical reality, in contrast to Lorentz who used it as mathematical formal method.

We can often see statement where was declared equivalence of gravitation and a non-inertial reference frame. On Figure 4 and 5 it is shown that the non-inertial reference frame or the accelerated reference system does not have "gravitational" field: it is beginning on and ending on bottom of the Einstein's elevator, or in other words, it does not exist. We observe gravitation phenomenon near of massive body.

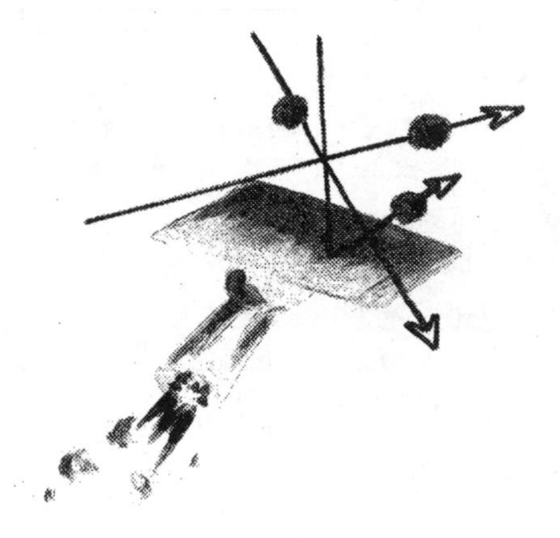

Figure 4. Thought experiment with bottom of the Einstein's elevator and test-mass, which demonstrate that we have no gravitation in accelerated reference system.

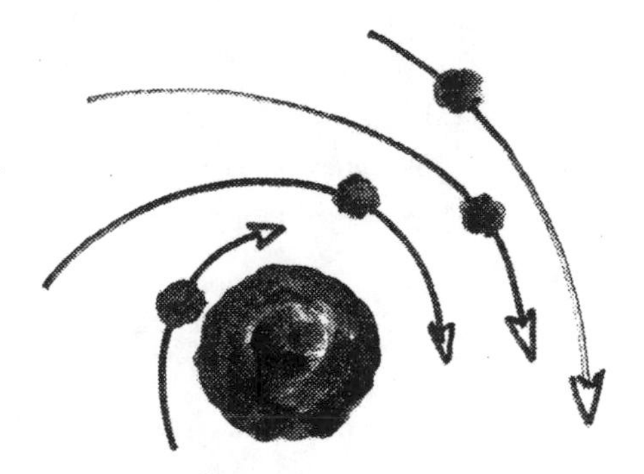

Figure 5. Gravitation as interaction on the distance massive body with test-mass.

In contemporary physics, which is based on the relativity theory, it strikes the eye, that, on the one hand, we have a lot loud noise about unification physics, and on the other hand, we have huge diversity theories, concepts – beginning from quantum scale and to cosmological theories – Lee Smolin argues unique Universe, and Max Tegmark and Andrei Linde develop multiple Universe. This diversity – per Lee Smolin [59], we have 10^{500} strings theories, can be taken as indirect evidence illogicality of the relativity theory – inconsistent theory permit everything.

Regarding to the relative theories we see problems with start or initial concepts as followed:

Declaration absolute velocity of light;

Inconsistency non-Euclid's geometry and it enforce on physical space;

Absence a physical evidence of interaction space and the Sun, space and light, or interaction space with then ever.

24. Physical method.

Physics was substituted by math so long time that was omit physical object – main subject of study. It seems that beginning of physics should start from reality, from physical objects and phenomena. Below we underline important, from our point of view, aspects of physical methodology.

1. Nature, observation of Nature, experiments with and observation of physical objects – all should be prior any notions, concepts or theories.

For preparation of observations and experiments should be used all previous knowledge, but should be open mind and critical thinking, for example,

if
you do not know what is light,
you know that ether is a artificial invention,
you make assumption about relation between light and ether,
you make assumption about relation between light and mirror,
you wait for particular picture on result of observation, because speed of the Earth is 30 km per second around the Sun and
you forgot that speed of the Sun system is 220 km per second around center of Milky Way,
and you do not get desired picture,
then
you do not have base to change fundamental paradigm on Nature and its properties,
because one, or a some combination or all your assumption, your inventions of notions or your expectations are wrong,
you should not conclude that
all your assumptions are wrong because you did not get desired picture and
contra assumptions were confirmed by this experiment.

2. Reproducibility. Often we see mention of five senses and intuition as sixth sense. Now we have more senses:

1) Sight;
2) Hearing;
3) Taste;
4) Smell;
5) Touch;
6) Proprioception – sense of space;
7) Time – biological clock;
8) Vestibular - Balance, equilibrioception;
9) Thermoception;
10) Nociception - physiological pain;
11) Descartes' intuition – clear mathematical intuition;

12) Undefined intuition – inside sense;
13) Undefined sense – old sixth sense.

Descartes' intuition usually is contrasted to other senses: all senses can and often wrong, but mathematical intuition never wrong. Descartes' or mathematical intuition is described as followed [4, page 91]:

"By intuition I [Descartes] understand not the unstable testimony of the senses, nor the deceptive judgment of the imagination with its useless constructions; but a conception of a pure and attentive mind so easy and so distinct that no doubt at all remains about that which we are understanding. Or, what amounts to the same thing, intuition is the undoubting conception of a pure and attentive mind, which comes from the light of reason alone and is more certain even than deduction because it is simpler; although, as we have noted above, the human mind cannot err in deduction, either. Thus everyone can see by intuition that he exists, that he thinks, that a triangle is bounded by only three lines, a sphere by a single surface, and other similar facts."

From historic point of view, all senses can make mistakes; include Descartes' intuition, imagination, the smartest mind and the biggest intellect. To be sure in correctness of our knowledge, we should get reproducibility of results with any combination our senses and power of mind, and different conditions of observation and experiments. Reproducibility is goal of science, but arts attempt to get unique result, and for religion enough to get the faith to dogma and speech of Authority.

2) Reflexivity. Physical reflectivity – study experiments by physical experiments. Study mathematical methods and they a range of applicability.
3) Self-definition self – identification: what is physics? Philosophy is generalizations of general and studies all; Mathematics is abstractions, idealizations, generalizations; Physics is study of the physical objects and phenomena, with mathematics and philosophy as tools.

Conclusions

All topics in this work are well known, they are not discovery of America. Here it is attempting to emphasize important, from our point of view, aspects of mathematics, physics and their interactions. Morris Kline had shown illogical development of mathematics in 1980. Here we discussed importance so called undefined or initial notions. Significance of initial notion cannot be overestimated. The inconsistency of initial concepts can ruin any theory.

The inconsistence of infinity was considered as initial notion set theory, also derived inconsistent statements of the set theory. It was shown that we can not operate any kind of infinite object, we cannot set up one-to-one correspondence for infinite set: when we try to do that we will get contradiction: in the same time part can equal to whole, and whole can equal itself with reference on the infinite one-to-one correspondence.

The signs of infinity and its algebra were brought to describe of the illusion of operation by infinite objects. Infinity is not number, but it is operated as number by particular algebra, where signs of infinity were mixed with numbers.

The newest contemporary theory of infinity – Sergeyev's Grossone Theory- was considered in detail and was shown that this theory brings new algebra and new inconsistencies.

The question 1 ? 0.999 … is well know and discussed thousand times in different works and places. Here we attempt to show that usage infinite expression like 0.999 … leads to opposite results, this usage and reference on the infinite expression should be excluded from logical reasoning.

The inconsistency of infinity close to initial notion of geometry: point, line, plain. We can see expressions: "infinite number of points on finite segment of line" and lose illogicality of words "infinite number." We discussed inconsistency of Euclid's, non-Euclid's geometry and the number theory.

The Banach-Tarski paradox, which often called as theorem, was present in work as proof of inconsistency topology by initial assumption of the infinite division of the finite geometrical object. We discuss also and other source of inconsistencies which mathematics brought to physics: negative and complex number, group theory.

The logic is borderline of science and religion activity, but it has problem too. Symbolic nature of mathematics leads to mutation of logic until acceptance complete illogicality, and this "new" logic is assigning to Nature.

Symbolic nature of mathematics and mathematical method as abstraction, idealization, generalization, - present oneself power and weakness of mathematics, and they was transferred to physics. There are inconsistent notions and theories, and reviving and reifying abstract notions.

To eliminate problems of usage of the symbolic mathematics, physics, on our point of view, should develop its own methodology based on reality Nature.

The first part of work was dedicated to discussing basically notions of mathematics. The second part of work was devoted in the main the notions of physics, in several cases with connection to mathematics. Here was tried implement physical methodology as initial consideration which beginning from physical object.

The standard definition of physics was presented to start discuss about initial notions of physics and still open physical questions.

From our point of view, first level of abstraction of the physical notions content notions substance, space and motion. Main properties of these notions are: conservation of substance and space, substance does not become space, and space does not become substance.

Next level of abstraction is the notions of dimensions: as three test body for usage of describing motion of other body.

The notion of mass still has diversity in its meaning, even inside the relativity theory. From our point of view, mass is absolute Newtonian notion in meaning of independency to inertial frame of reference.

The physical meaning of notion of time was represented as standard body with uniform periodic motion, and also independent to described motion. Mix abstract notions dimensions and time leads to illogicality from physical and mathematical points of view.

The Maxwell's equations still have several not solved problems, but they still are used as basement of description of physical phenomena.

The problems of Maxwell's equations lead to still open question: what is light? It is one of the main initial problems of physics.

The Maxwell's equations problems derived from unsolved problems of notion of electric charge. We have problems with measuring of electrostatic charges, and description of electrostatic phenomena. "Successfully" can be described the separate taken one electron only.

The notion of electron has diversity of models and mathematical descriptions. It is second of main initial physical problem: what is electron? It looks like combination of abstract notions atom and charge without appropriate physical object.

Several problems of the relativity theory was emphasized with connection to initial notion the absolute velocity of light, inconsistency of non-Euclid's geometry, and experiments with deviation of light.

Historical perspectives of evolution of the man's thoughts are:

Speaking man

Man speaks to stone, wind, rain, Sun ...

Man's religions

Man's philosophy

Man's science, separate fields

Anthropic principle: Universe or multiple Universes speak to man:

I'm existing or we are existing for you, because you!

It is, from our point of view, the high level of the man's arrogance.

Alternative approach is:

Exist: Nature, Universe as something fare, and a physical reality as something close.

Do not Exist in reality words (and their meaning) of the symbolic language in wide sense: everyday language, math language, science language, notions, concepts, theories, symbols and so on, all of those exist in man's mind and on the paper. Words and their meaning are related to or about reality.

Nature does not have and does not need any laws, theories, languages, all of those are needs of men for accumulate experience, for compress knowledge, for transfer experience to next generations.

To fall on physicist's head apple does not need to know Newton's laws, to swim fish does not need to know the Navier–Stokes equations, to fly seagull does not need to know aerodynamics.

All **symbolic languages** do not exist in reality because all notions and concepts are abstract and consist inside contradictions. Ideal abstraction cannot exist even on paper, and can exist in man's mind only. Euclid's Point without parts, line as length alone, a two dimension plane exist on paper as a rough physical model, point on paper is a huge number of molecules of ink, paint, or graphite. If you get a book written on unknown for you language then the book's information does not exist for you. Geometrical shapes are international, but it means that all people have a similar ability to make abstraction, and they observe the same Nature, and have the same result of abstraction. As was shown, the ideal abstract geometrical objects related to solid body and exist in man's mind only.

It is position against of Weinberg's philosophy: ingredients of any theory do not exist in reality. And more: notions and theories have inconsistencies and illogicality, but Nature doesn't. Man is so weak that prefer combine contradictions in special logic – dialectic, then attempt assign those contradictions to Nature, but Nature does not have contradictions and inconsistencies.

Hegel (1770-1831) had noticed that we cannot eat fruits, but we can eat apple, cherry and so on. Very easy to note that we cannot eat apple too – I can eat only the Cortland red apple which I keep in my left arm.

From time of the Aristotle's Physics, the second reborn of physics was made by Galileo and Newton per combination of mathematics, natural philosophy and experiments with observations. Long time of usage symbols in math and physics brought to us habit to correspond notion and real object or phenomenon. Next step per habit follows creation new notion and the faith that real object, which correspond to this new notion, should exist. Repeating those steps did create big problem for mathematics and physics:

In math are used inconsistent abstract notions,

In physics were and are "discovered" abstract notions which exist in man mind only as real object,

And the biggest physical problem is the experimental confirmation of existence of abstract notion, especially, existence of the inconsistent notions.

Lack of reflection in math show us that mathematicians "invented" methods to hide – not solve – this problem. There are: acceptance so called "undefined" notions, axiomatic method, acceptance conflicting sets of axioms, and so on.

Lack of reflection in physics brings the complete faith to math; loosing physical objects, transfer an inconsistency of man's theory to reality, declaration and the faith that Nature follows the math laws, experimental confirmation of existence of a non-existent.

The problems of physics are connected to a very broad range of issues. We have at least two written evidence of mental problem of physicists. First one is book by Emanuel Derman [60], "My Life as a Quant. Reflection on Physics and Finance," 2004. Second one is book by Lee Smolin [59], "The Trouble with Physics. The Rise of

String Theory, the Fall of a Science, and What Comes Next," 2006. Emanuel Derman had tried to meditate but had left the physics and gone to finance. Philosopher Paul Feyerabend had helped Smolin to stay in physics.

Community of contemporary physicists is closed. Description of the closed community of physicists we can see in works by Steven Weinberg, Lee Smolin, and Lev Okun. Lee Smolin gives a convincing classical definition of a closed community of contemporary physicists, in 7 items [59, p.284]. Smolin sets forth the requirements of an open community too, in 6 items [59, pp. 301, 302], with two main criteria: peer review, and "allegiance and continued adherence to the shared ethic." He described problem of the physical education as approach - "Shut up and calculate" [59, 61].

Here it must be emphasized, that the physics occupies a special place - connection to the military field, atomic weapons, and even to political and economical "weapons." In time the Cold War between USA and USSR, construction of super collider was economical problem for USSR. After the Cold War both of country had stopped construction of the super colliders. Now USA and Russia do not participate in financing of European Supercollider, only American and Russian physicists work in CERN.

From the time of the Morris Kline's book in 1980 gone 34 years, and we cannot see even tunnel, where we could hope to see light of scientific physics. It looks like everybody prefers enjoy the pseudoscientific fiction at public account.

Acknowledgments

I'd like to thank Dennis Allen and Gregory Volk for discussions and preparation this work. Special thanks to Eugene Shklyar and Sergey Arteha for discussions and constant support my works.

Appendix. Review of the book's topics per web site edge.com.

On web site [62] each year from 1998 put question and contributors write their answers.

The Edge Question 2014 is : WHAT SCIENTIFIC IDEA IS READY FOR RETIREMENT?

Contributors - 177, All Responses - 175, Ideas - 181.

Review of responses shown broad understanding of "retirement of idea", and proposed idea in essay looks like particular meaning of idea in especial conditions.

Richard H. Thaler say: "Don't Give Wrong Theories Funerals. Just Stop Treating Them as True."

Just Stop Treating illogical concepts and theories as True - is proposition this book too.

Ian McEwan proposed: "Beware of arrogance! Retire nothing!"

Carlo Rovelli proposed to retire notion of "geometry", but in his comment it means quantum geometry,

which in quantum scale even does not exist by Carlo Rovelli, if so – nothing to retire.

First of all, theme of infinity flies in air.

Max Tegmark, (Physicist, MIT; Researcher, Precision Cosmology; Scientific Director, Foundational Questions Institute), proposed to retire "INFINITY".

But it seems that Tegmark did not read Weil, Gilbert, or Eli Maor, and a lot other works, where we can find definition of mathematics as science about infinity. If it is right then we, and Max Tegmark too, should retire mathematics too.

In this case book by Max Tegmark [63], "Our Mathematical Universe: My Quest for the Ultimate Nature of Reality", 2014, where he is "arguing that mathematics describes the universe so well because the universe ultimately is mathematics", should be retired too.

Gregory Benford proposed to retire "The Intrinsic Beauty and Elegance Of Mathematics Allows It to Describe Nature."

Problems with contemporary physics are illustrated a lot propositions to retire the Weinberg's final theory.

There are propositions to retire:

Geoffrey West	- Theory of Everything;
Eric R. Weinstein	- M-theory / String Theory is the Only Game in Town ;
Frank Tipler	- String Theory;
Paul Steinhardt	- Theories of Anything.

There are propositions against logic to retire concepts:

Eldar Shafir	- Opposites Can't Both Be Right;
Robert Provine	- Common Sense;
Alan Alda	- Things Are Either True Or False.

There are propositions to retire science and its main concepts:

Ian Bogost	- "Science";
Alex Holcombe	- Science Is Self-Correcting;
Aubrey De Grey	- Science Progresses Most Efficiently By Allocating Funds Via Peer Review;
Sam Harris	- Our Narrow Definition of "Science";
George Dyson	- Science *and* Technology;
Victoria Stodden	- Reproducibility;
Sean Carroll	- Falsifiability.

References

[1] Kline Morris, Mathematics and The Physical World, 1959, Dover Publications, Inc. New York, 483pp.

[2] Kline Morris, Mathematics. The Loss of Certainty, 1980, Barnes & Noble, New York, 448pp.

[3] Kline Morris, Mathematical Thought from Ancient to Modern Times, OXFORD UNIVERSITY PRESS, New York, 1972, 1238 pp.

[4] Kline Morris, Mathematics and the Search for Knowledge, OXFORD UNIVERSITY PRESS, New York, 1985, 257pp.

[5] Boyer, Carl Benjamin, History of Mathematics, Second Edition, JOHN WILEY & SONS, INC., New York, 1991, 715 pp.

[6] Katz, Victor J., A History of Mathematics. An Introduction. Harper Collins College Publishers, New York, 1993, 786 pp.

[7] V. Kanke. Philosophy of mathematics, physics, chemistry, biology: tutorial. (RUSSIAN) Канке В.А., Философия математики, физики, химии, биологии: учебное пособие. М.: КНОРУС, 2011.

[8] The Story of 1. Nick Murphy, Terry Jones. BBC. 2005.

[9] Maxwell J. An Elementary Treatise on Electricity. Second Edition, 1888, 2005. DOVER PUBLICATION, INC. Mineola, New York. 234p.

[10] E. Wigner, "The Unreasonable Effectiveness of Mathematics in the Natural Science," *Comm. Pure & App. Math.* **13** (1): 1-14 (Feb 1960).

[11] R. Penrose, The Road to Reality: A Complete Guide to the Laws of the Universe. Alfred A. Knopf, 2004, 1099p.

[12] G. Ivchenkov, The Most Important Confirmation of GRT or What Did Lord Eddington Measure in 1919? (in Russian), http://www. elibrary-antidogma.narod.ru/bibliography/eddington.htm.

[13] Landau L.D., Lifshitz E.M., The Classical Theory of Fields. Vol. 2. 2002, pp. 428.

[14] W. Heisenberg, The Physical Principles of The Quantum Theory (Dover, 1949).

[15] Rucker, Rudy, INFINITY and the MIND. The Science and Philosophy of the Infinite. PRINCETON UNIVERSITY PRESS, PRINCETON, NEW JERSEY, 1995, 342 pp.

[16] Maor, Eli, To Infinity and Beyond. A Cultural History of the Infinite., Birkhauser, Boston, 1987, 275 pp.

[17] Amir D. Aczel, The Mystery of the Aleph. Mathemetics, the Kabbalah, and the Search for Infinity. WASHINGTON SQUARE PRESS, New York, 2000, 258 pp.

[18] Wallace, David Foster. Everything and More: a compact history of infinity. 2003, W.W. Norton & Company, Inc. NY. 344pp.

[19] David Tall, Intuition of Infinity, Mathematics in school, 1981.

[20] Vilenkin N. The search for infinity. (Russian) Виленкин Н. Я. В поисках бесконечности.— М.: Наука, 1983. 160 с.

[21] Encyclopedic Dictionary of Mathematics. Second Edition. Vol. 2. The MIT Press 1996, pp. 2148.

[22] Anti-Dühring by Frederick Engels, 1877.

[23] A. D. Aleksandrov, A. N. Kolmogorov, M. A. Lavrent'ev, Eds., Mathematics: Its Content, Methods, and Meaning (Dover, 1999).

[24] Sergeyev Y. Solving ordinary differential equations on the Infinity Computer by working with infinitesimals numerically. 2013. http://si.deis.unical.it/~yaro/ODE.pdf.

[25] http://si.deis.unical.it/~yaro/arithmetic.html.

[26] Sergeyev Y. NUMERICAL COMPUTATIONS AND MATHEMATICAL MODELLING WITH INFINITE AND INFINITESIMAL NUMBERS. 2009. http://si.deis.unical.it/~yaro/Calcolo.pdf

[27] Sergeyev Y. A NEW APPLIED APPROACH FOR EXECUTING COMPUTATIONS WITH INFINITE AND INFINITESIMAL QUANTITIES. 2008. http://si.deis.unical.it/~yaro/Numerical.pdf

[28] http://en.wikipedia.org/wiki/0.999 ...

[29] THE THIRTEEN BOOKS OF EUCLID'S ELEMENTS. TRANSLATED FROM THE TEXT OF HEIBERG WITH INTRODUCTION AND COMMENTARY BY SIR THOMAS L. HEATH. Second Edition revised with additions. Volume I, Introduction and Books I, II. DOVER PUBLICATIONS, INC. New York. 1956. 432 pp.

[30] http://en.wikipedia.org/wiki/Banach%E2%80%93Tarski_paradox

[31] Vladimirov Y. Metaphysics. M. 2009. 568p. (Russian).

[32] Physical Encyclopedia. Editor A. Prokhorov. Moscow. 1988.

[33] Karl R. Popper, Conjectures and Refutations: The Growth of Scientific Knowledge, 1968, Harper & Row, Publishers, Inc., 417 p. page 312. What is Dialectic?

[34] G. Frege, The Foundations of Arithmetic (Northwestern University Press, 1996).

[35] Steven Weinberg, *Dreams of a Final Theory: The Search for the Fundamental Laws of Nature* (1993), 340pp.

[36] [http://en.wikipedia.org/wiki/Thermodynamics] [Sir William Thomson, LL.D. D.C.L., F.R.S. (1882). Mathematical and Physical Papers 1. London, Cambridge: C.J. Clay, M.A. & Son, Cambridge University Press. p. 232.]

[37] S. Hawking, The Universe in a Nutshell (Bantam Books, 2001)

[38] The American Heritage® Dictionary of the English Language, Fourth Edition copyright ©2000 by Houghton Mifflin Company. Updated in 2009. Published by Houghton Mifflin Company. All rights reserved.

[39] PHYSICS. PRINCIPLES WITH APPLICATION. Fifth Edition. Douglas C. Giancoli. 1998. Prentice Hall, Upper Saddle River, New Jersey 07458. 1096 pp. http://www.prenhall.com/giancoli

[40] http://en.wikipedia.org/wiki/Physics

[41] Whitrow G.J. The Natural Philosophy Of Time. Oxford.1980. 399pp.

[42] Reichenbach H. The Philosophy Of Space & Time. Dover. 1958. 295pp.

[43] Time, Reality& Experience. Ed. By Callender C. Canbridge. 2002. 328pp.

[44] Barbour J. The End Of Time. Oxford. 1999. 374pp.

[45] A Two-Time Universe? USC College physicist explores how a second dimension of time could unify physics laws, better describe the natural world. http://www.usc.edu/schools/college/news/may_2007/bars.html.

[46] D. Pavlov. Four-dimensional time as alternative to Minkowski's space-time. (Russian) Д. Г. Павлов. ЧЕТЫРЁХМЕРНОЕ ВРЕМЯ КАК АЛЬТЕРНАТИВА ПРОСТРАНСТВУ-ВРЕМЕНИ МИНКОВСКОГО. Московский Государственный Технический Университет им. Н.Э. Баумана, НИИЭМ, Москва, Россия. http://www.chronos.msu.ru/RREPORTS/pavlov_chetyrekhmernoe/pavlov_chetyrekhmernoe.htm.

[47] Lee Smolin. Time Reborn. From the Crisis in Physics to the Future of the Universe. 2013, Spin Networks, Ltd.

[48] Levich A.P. Generating Flows and a Substantional Model of Space-Time //Gravitation and Cosmology. 1995c. V.1.

№3. Pp. 237-242. http://www.chronos.msu.ru/EREPORTS/levich_generat.flows.htm

[49] J. C. Maxwell, A Treatise on Electricity and Magnetism, Vol 1 (Dover, 1954). 506 p.

[50] J. C. Maxwell, A Treatise on Electricity and Magnetism, Vol 2 (Dover, 1954). 500 p.

[51] Fock V. Theory of space, time and gravitation.(into Russian). Фок В. Теория пространства, времени и тяготения. Москва. 1961.563с.

[52] McMahon D. Relativity demystified. McGRAW-HILL, New York 2006. 344p.

[53] Lev B. Okun, Energy and Mass in Relativity Theory. World Scientific Publishing Co. Pte. Ltd., 2009, 311pp. (30 Okun's works).

[54] Khrapko R. What is mass? UFN. (Russian) Р.И.Храпко, Что есть масса?, УФН, т.170,№ 12, 2000 г. стр. 1363-1366, библ. 11 назв.

[55] Max Jammer. Concepts of Mass in Contemporary Physics and Philosophy. PRINSTON UNIVERSITY PRESS. Princeton, New Jersey. 2000.

[56] http://www.antidogma.ru/index_ru.html

[57] http://www.worldnpa.org/site/

[58] V. Ginzburg. How and who create the special relativity theory. (Russian) В. Л. Гинзбург. Как и кто создал специальную теорию относительности? В книге В. А. Угаров. Специальная теория относительности. М., 1977., 384 стр.

[59] L. Smolin, The Trouble with Physics. The Rise of String Theory, the Fall of a Science, and What Comes Next (Houghton Mifflin, 2006).

[60] Emanuel Derman. My Life as a Quant. Reflection on Physics and Finance. JOHN WILEY & SONS, INC. 2004.

[61] David Mermin,"Could Feynman Have Said This?," Physics Today (May 2004), p. 10.

[62] http://edge.org/responses/what-scientific-idea-is-ready-for-retirement,

[63] Max Tegmark, Our Mathematical Universe: My Quest for the Ultimate Nature of Reality, Alfred A. Knopf, New York, 2014, 421 pages.

Name Index

Anaximander (610 – 546 BC)
Parmenides (510-450BC)
Zeno of Elea (490 – 430 BC)
Aristotle (384 BCE – 322 BCE)
Archimedes (270-212BC)
Titus Lucretius (99 BC – 55 BC)
Nicolaus Copernicus (1473 – 1543)
Galileo Galilei (1564 – 1642).
René Descartes (1596 –1650)
Newton, Isaac (1642 – 1727).
Georg Hegel (1770 –1831)
Julius Dedekind (1831 –1916)
Georg Cantor (1845 –1918).
John McTaggart (1866 – 1925)
Ernest Rutherford (1871 – 1937)
Godfrey Hardy (1877 – 1947)
Albert Einstein (1879 – 1955)
Arthur Eddington (1882 – 1944)
Niels Bohr (1885 – 1962).
Hermann Weyl (1885-1955)
Vladimir Fock (1898 –1974).
Eugene Wigner (1902 –1995)
Carl Boyer (1906 –1976)
Morris Kline (1908 – 1992)
Vitaly Ginzburg (1916 –2009)
Richard Feynman (1918 – 1988)
Lev Okun (1929 – present)
Steven Weinberg, (1933 - present)
Yuri Vladimirov (1938 – present)
Rucker, Rudy (1946 – present)
Amir Aczel (1950 - present)
Lee Smolin (1955-present)
David Wallace (1962 – 2008)
Sergeyev Yaroslav (1963 - present)
Max Tegmark (1967-present)

About the Book

Well known that mathematics and physics have problems in their development. Only one mathematician, Morris Kline, discovered illogicality of development of mathematics. Despite this, he attempted to justify illogicality in math by fruitfulness of usage of mathematics in physics, instead to stay problem about illogical development of physics.

Here is discussing inconsistencies of undefined notions which are reasons of paradoxes. Main initial notion of mathematics is notion of infinity, and it has inconsistence and this inconsistency is distributed to derived notions of infinitesimal and continuity. Those notions related to almost all branches of mathematics which used physics.

Also in work is considering miss inconsistencies of Euclid's and non-Euclid's geometries. A lot approaches like "physics is geometry or geometry is physics" was and is ignoring those inconsistencies of geometries.